Praise for *T*

"Clear-eyed, crisply written . . . [*T*] entertains as it educates . . . [and Hooven] circles back to her own story, speaking to us intimately, casting her intellectual journey in a luminous chiaroscuro. . . . Her book confronts ugly truths about male behavior, but also seeks to reintroduce nuance into our discourse by enlarging our grasp of the biological processes shaped by testosterone. *T* is a gorgeous culmination of an odyssey both professional and personal."

—*The Wall Street Journal*

"Hooven makes a compelling case that T is a powerful influence on our bodies and brains. . . . As the final chapter of [*T*] argues, it's hard to make a start on . . . social improvements if we don't fully understand why things are the way they are. Clear-eyed books like this, which mercifully avoid culture-war partisanship, are a great start on that particular quest." —*The Times* (UK)

"One of the most compelling books on human behavior I've ever read, *T* is a scientific mystery story told with insight, intelligence, and panache."

—**Daniel Gilbert, Edgar Pierce Professor of Psychology at Harvard University and author of the *New York Times* bestseller *Stumbling on Happiness***

"Hooven's review of what testosterone does and does not do is science-writing at its best: intriguing, personal, bold, persuasive, and most importantly, transparent. Her gripping account will fascinate, whether you're a teenager in the throes of puberty or are just curious about the nature of sex and gender—one of the most important debates of our time."

—**Richard Wrangham, author of *The Goodness Paradox***

"With all the talk about testosterone in sex, sports, and politics, we need a good explanation of the science and its implications, and this one is outstanding." —**Steven Pinker, Johnstone Family Professor of Psychology at Harvard University and author of *How the Mind Works* and *Enlightenment Now***

"Filled with fascinating—often astounding—insights, *T* takes us on a journey from the discovery of this powerful hormone, to its effects throughout the animal kingdom, to a crucial understanding of how it shapes human minds, bodies, and behavior. Hooven is a compelling and compassionate guide, and *T* is a book we need now."

—**Emily Yoffe, journalist and former Dear Prudence columnist at *Slate***

"*T* is the story of the most famous, most misunderstood, and most maligned chemical in our bodies: testosterone. A fascinating, brave, and brilliant book—the best I've read on the topic."

—**Steve Stewart-Williams, author of *The Ape That Understood the Universe***

"A superb and engaging book that delivers the unfiltered truth about testosterone, sex, and sex differences, told with clarity and compassion. *T* conveys a deep understanding about the hidden power of testosterone in our lives, but also teaches us how scientific knowledge exposes harmful misconceptions and helps us become more humane."

—**Daniel E. Lieberman, author of *Exercised: Why Something We Never Evolved to Do Is Healthy and Rewarding***

"With wit, warmth, and a touch of motherly love, Hooven lucidly lays out a formidable scientific case for how and why the sexes are different. Boldly confronting contemporary gender issues, *T* speaks directly to why getting human nature right matters for making the world a better place." —**Joe Henrich, professor of human evolutionary biology at Harvard University and author of *The WEIRDest People in the World* and *The Secret of Our Success***

"Dr. Carole Hooven has written the definitive book on testosterone . . . [which] examines testosterone from every possible angle. . . . If we can educate ourselves about 'T,' according to Hooven, then the unfiltered truth about the impact of testosterone and the differences between the sexes will add to our knowledge-base and help us create a better society." —*Evening Standard*

"We need to be more accepting. . . . Human beings are slow to learn, and it takes us a long time to adopt all this information and adapt. . . . We need more examples like you. You're raw, and you're open, and you're confident enough to be vulnerable. . . . It was a great conversation. I've never cried more." —**Joe Rogan, *The Joe Rogan Experience***

"Truly one of the most remarkable and fascinating books that I've come across in a long time. . . . Carole is that rare scientist who is able to take sometimes complicated and technical information and make it all understandable."
  —**Danielle Crittenden, *The Femsplainers Podcast***

"This is a smart, balanced, interesting, humane, and rigorous book that pulls no punches. It helps us understand better the relations between men and women, homosexuality, transgenderism—all aspects of our lives." —**Andrew Sullivan, *The Weekly Dish***

"It's a good narrative, not just an academic book, but I like that you get into the nitty-gritty of the science."
  —**Razib Khan, *Unsupervised Learning***

"Timely and intriguing . . . [*T* investigates] the ancient question of nature vs. nurture, and Hooven walks a fine line between the two. . . . Popular science likely to excite academic debate on sex and gender."
  —*Kirkus Reviews*

"Hooven separates testosterone fact from fiction in her comprehensive debut. . . . She takes readers to high-tech labs and on a day of 'chimping' in Uganda's Kibale forest—and her writing is refreshingly free of jargon. The result is an approachable introduction to an often misunderstood aspect of human biology." —*Publishers Weekly*

"[Hooven] deftly details the tremendous impact of testosterone, . . . [and t]he complex interplay between genes, hormones, social environment, and experience makes us who we are. . . . Hooven effectively counters erroneous or misleading assertions about testosterone, a hormone so culturally familiar to most of us that it's frequently designated by a single letter." —*Booklist*

"[Hooven's] insight into evolutionary biology and human behavior will be valuable not only to scientists but to anyone interested in social change. . . . A provocative, academic, accessible look at the science behind human behavior." —*Library Journal*

"Hooven makes her case for the importance of T with a humility, humor, and grace not often seen in our cultural battles these days."
—**Stephen L. Carter,** *Bloomberg Opinion*

The Story of TESTOSTERONE,
the HORMONE that
DOMINATES and DIVIDES Us

CAROLE HOOVEN

A HOLT PAPERBACK   HENRY HOLT AND COMPANY   NEW YORK

Holt Paperbacks
Henry Holt and Company
Publishers since 1866
120 Broadway
New York, New York 10271
www.henryholt.com

A Holt Paperback® and ⒽⒹ® are registered trademarks of
Macmillan Publishing Group, LLC.

The Library of Congress has cataloged the hardcover edition as follows:

Names: Hooven, Carole, author.
Title: T : the story of testosterone, the hormone that dominates and
  divides us / Carole Hooven, PhD.
Description: First US edition. | New York, New York : Henry Holt and
  Company, 2021. | Includes bibliographical references and index.
Identifiers: LCCN 2020051032 (print) | LCCN 2020051033 (ebook) |
  ISBN 9781250236067 (hardcover) | ISBN 9781250236074 (ebook)
Subjects: LCSH: Testosterone. | Testosterone—Physiology.
Classification: LCC QP572.T4 H66 2021 (print) | LCC QP572.T4 (ebook) |
  DDC 612.6/14—dc23
LC record available at https://lccn.loc.gov/2020051032
LC ebook record available at https://lccn.loc.gov/2020051033

ISBN: 9781250811455 (trade paperback)

Our books may be purchased in bulk for promotional, educational, or business use. Please
contact your local bookseller or the Macmillan Corporate and Premium Sales Department at
(800) 221-7945, extension 5442, or by e-mail at MacmillanSpecialMarkets@macmillan.com.

Originally published in hardcover in 2021 by Henry Holt and Company

First Holt Paperbacks Edition 2022

Designed by Kelly S. Too

Printed in the United States of America

1 3 5 7 9 10 8 6 4 2

For Griffin

CONTENTS

1

STARTING OUT

CHIMPING

If we were not under their sleeping nests when the chimpanzees woke up, then we would miss out on the pee—the key to the chimps' testosterone levels. So I prepared, as I did almost every morning during the eight months I spent with the chimps, for a walk through the predawn jungle.

Evolution has fashioned an elegant system that motivates us to start our days with the benefit of the light (and heat) emitted by our star. Like all diurnal animals (those that are active during the day), we sync our sleep-wake cycles with our planet's twenty-four-hour rotation around its axis. When the morning sun is sensed by the photoreceptor cells in our retinas, the information is transmitted to the pineal gland, a tiny pine cone–shaped organ, deep in the center of our brain. In response, the gland decreases its production of the "sleep hormone" melatonin, which nudges us toward a particular behavior—waking up.

At least, this is how it worked before we humans got used to artificial light. But since chimps are sticking to the old schedule, I had to get my butt out of bed while my melatonin was still high. I attempted to counteract this groggy state of affairs with a dose of caffeine delivered via coffee, which I made with rainwater on the field site's propane-powered stove.

Armed with my Wellies to protect from such inconveniences as army ants, mud holes, and black mamba snakes, a flashlight, and a foot-long machete (for bushwhacking), I headed out to meet my Ugandan field assistants. This was a usual day of chimping—tagging along with the chimps and taking notes on their lives and activities in the Kibale Forest of Western Uganda.

After a hike of about an hour, I rested on the forest floor near the base of one of the trees in which the chimps slept, high in the treetop nests they'd constructed the night before. I tried to soak up every detail of the dramatic transformation of the night forest. The steady hum of trilling insects was drowned out by the growing cacophony of bird and monkey calls, and slivers of sunlight pierced through the undergrowth, turning dewdrops into glowing gold beads that clung to the green foliage. I was waiting for one sound in particular, that of rustling from above, the first stirrings of chimps waking up. That was my cue to prepare.

Chimps aren't much different from humans in terms of their first morning needs—they need to go! But while we stumble out of bed and head to the bathroom (or outhouse or pit), chimps just stick their rear ends out, over the side of their nest. I did my best (which was not always good enough) to be far enough away to shield myself from the urine that rained down about thirty feet through the leaves, but close enough that I could catch some. I did that with the help of a long stick with a forked end, over which I had tied a plastic bag.

In this way I was making a small contribution to the data, both behavioral and physiological, that was collected by researchers from the Kibale Chimpanzee Project. This treasure trove of information allows scientists to gain insights into the origins of all kinds of behaviors. But we were particularly interested in sex, aggression, and dominance, which are all affected by the subject of this book: testosterone, or, as insiders call it, "T." With human subjects, we can just ask them to spit into a tube. But wild chimps are less cooperative, so we measure T in their urine (and feces) instead.

I carefully pipetted what little pee I managed to catch from the plastic bag into test tubes, to be carried back to the field site for later transport back to the endocrinology lab at Harvard. After a few minutes of rustling and relieving, the chimps shimmied down the tree trunks to begin their day, with the field assistants and me in tow.

A BULLY GIVES A BEATING

Chimpanzees live in "communities," usually comprising about fifty chimps. In some ways, each community is like a group of people living in a small town, with well-defined and defended borders, and hostile relationships with the neighboring towns. Imoso was like the mayor—the alpha—of this town, called Kanyawara. It was just one of several such communities located within the vast forest near the border with the Democratic Republic of the Congo. Imoso was temperamental and despotic—a leader who must have been more feared than liked. Each day, smaller groups of chimps from the community, called "parties," would congregate and spend the day together, and I'd follow one of the parties. When Imoso was part of a party I followed, I could count on lots of grunting, screaming and hooting, threatening, slapping, stick dragging and throwing, and chest pounding. There was one sure way to make things even more exciting, and that was to add a fertile (estrous) female to the mix. Lots of sex, and even more aggression, as the males competed for the right to mate with her.

Other days were punctuated less with bouts of high drama, and more with the ebb and flow of nurturance and play. Little ones would cling and cuddle, nurse, tumble and chase their siblings and friends, or perch like royalty on mom's back, as they traveled from one feeding patch to another. Those were the times I followed parties without the adult males.

On one January day, Imoso seemed calmer than usual. And on this day, unusually, he decided he'd like to hang out with only one female and her two small children. Leaning against a tall fig tree, I opened my notebook. Outamba sat behind Imoso on a large downed tree in a clearing. She expertly searched through his thick, dark hair, separating and flattening it, examining the area for dirt or parasites, deftly removing what she found and popping the tasty bits into her mouth. The baby Kilimi and her older sister Tenkere frolicked in a patch of grass in the heat of the equatorial midday sun, amid the din of birds and insects.

Outamba's ear-piercing screams jolted me out of my tranquil state and sent my heart racing. I bolted upright. Imoso jumped up so that he stood on the fallen tree and began to pound Outamba with his fists while also kicking her. She tumbled to the ground, and tiny Kilimi quickly hopped into the refuge of her arms. Outamba huddled over her daughter in a protective embrace, her back exposed to Imoso's assault. I tried to accurately record all that was happening—who was doing what to whom, and for exactly how long. (I was lucky to have been with one of the project's experienced field assistants, John Barwogeza, who gave me a thorough debriefing on all that I had missed.) After a few minutes of this brutality, already the longest and most severe beating I'd ever seen, Imoso picked up a large stick and began hitting Outamba's head and back with it. Tenkere, just three years old and no more than two feet tall, raced around Imoso and pounded at him with her impotent little fists as the giant beat up her mother. But kicking, hitting with fists and sticks wasn't enough—Imoso got even more creative and hung from a branch, leaving his feet free to stomp on and kick her with greater force. Nine stunning minutes later, it was over.

The beating left Outamba bleeding from the tender, hairless skin on her bottom, but at least her kids were unharmed and she was able to scamper off with them.

Although I knew other researchers had observed prolonged—even murderous—attacks, this was new for me. The episode was gut-wrenching, but also, as a scientist, it was thrilling and confusing. Sure, the big males would routinely harass and beat the adult females, but from what I'd seen before, these beatings were brief and mild by comparison.

Richard Wrangham, the world-renowned Harvard primatologist who founded and ran the field site, happened to be visiting that week. I raced the two or so miles back through the forest to the field station to describe what I'd seen. I was breathless and bursting with emotion and questions, but his initial response was to simply hold out his hand to shake mine. He told me that I was the first researcher to observe such weapon use by a nonhuman primate in the wild. *Time* magazine even ran a story, accompanied by a large picture of Richard, me, and the now famous stick (later retrieved from the clearing by the field assistants), under the title "Wife Beaters of Kibale." That anthropomorphic title made me cringe, but there was no denying the similarities between Imoso's disturbing behavior and domestic violence among humans. Why did he do it? I didn't have any answers that day, but research on testosterone and reproduction from the field site would later supply them.

DEMONIC MALES

My journey to Uganda wasn't exactly a straight line. An interest in human behavior drove me to major in psychology in college. I enjoyed classes like Freud and Jung, Abnormal Psychology, and Personality and Individual Differences. But it wasn't until my senior year that I had to restrain myself from jumping out of my seat, barely able to contain my excitement about the lecture material. I will never forget that course (Biological Psychology), the professor (Josephine Wilson), and the day she introduced me to neurons and neurotransmitters and how their actions and levels affect all kinds of behavior. I remember her standing tall, raising her outstretched arms above her head, and wiggling her fingers around to bring a neuron and its dendrites—little branches that communicate with other neurons—to life. A new, powerful way to understand the origins of behavior was opening up for me, and it felt tremendously satisfying. I knew I wanted more of that feeling, but graduation loomed and I had no job.

As one does with a BA in psychology, I landed a job in financial software. (Mostly I just wanted a job where I could "work with computers." This was 1988, after all.) I told myself that I would do this for a couple of years until I figured out a grand life plan. But I still had lots of learning to do, and my job was comfortable. So two years turned into ten. I took classes that I had missed out on in college, like molecular biology and genetics, and discovered that—contrary to the impression I'd gotten throughout my early schooling—I loved biology. I traveled widely, to places like Israel, Tanzania, Costa Rica, and China, and became curious about the origins of the diversity of cultures and ecologies across the globe. And I read popular science books, like *The Selfish Gene* by Richard Dawkins, that introduced me to how evolutionary theory could help to answer my questions about life on earth.

These experiences intensified my desire to find the deepest, most powerful explanations for human behavior and converged on one question: how has evolution shaped human nature?

Then I read the book that suggested a path I could take to pursue my questions: *Demonic Males: Apes and the Origins of Human Violence.* It wasn't violence, specifically, that drew me in; it was the approach that the two authors used to investigate large questions about how we got to be the way we are. I decided that I wanted to do what the lead author did, which was to study chimpanzees to learn more about ourselves and our evolutionary origins. So I quit my job and applied to graduate school.

I do not recommend doing things in that order.

The lead author of that book was Richard Wrangham. Fortunately he happened to teach at Harvard, in my home city of Cambridge, Massachusetts. I eagerly mailed in my application to the program in his department, which at the time was called Biological Anthropology. The rejection note was disappointing, but in hindsight I should have expected it. It's tough to get into such a program without a shred of research experience "in the field," as it's called. But in some cases, naivete can be an asset. I persisted in my efforts, and eventually Richard—we were now on a first-name basis—offered me an opportunity to spend a year in Uganda at the Kibale Chimpanzee Project. He had started the field site in 1987 to study the behavior, physiology, and ecology of wild chimpanzees. My job was to manage the site and learn to conduct some research of my own. I could hardly believe it. Of course I accepted.

SEX AND VIOLENCE IN TWO KINDS OF PRIMATES

And that's how I found myself in the forest on that January day in 1999, catching chimp urine and watching a big male beat up a smaller female while she tried to protect her kids. Their interaction had dramatically exemplified the contrasting patterns of chimp behavior that had already captivated my interest—relatively peaceful, nurturing females and sex- and hierarchy-obsessed aggressive males.

I watched adult male chimpanzees use aggression in different situations for different purposes, only some of which had a clear explanation. They used it to show who was in charge and to demand what we might think of as respect. A lack of respect meant that one's dominance rank was not being observed, and a beating might be the way to increase the chances that proper deference be paid to the dominant male in the future. Two males, close in dominance rank, might duke it out over a sexual opportunity—fighting to win a mating with a sexually attractive female (those who are in estrus and who can thus conceive are the focus of much male attention), or to keep other males away from her (known as "mate guarding"). And what about Imoso's assault on Outamba, at a time when she wasn't in estrus? As the data would later suggest, such aggression tends to increase a female's sexual compliance in the future. Males tend to target females who are in the best reproductive condition, and females preferentially mate with and bear offspring fathered by males who have been especially aggressive toward them. (I should emphasize that this doesn't mean that men's aggression against women has a similar evolutionary rationale, still less that such behavior is inevitable or excusable. And in any case, other animals—including other primates with different social systems—can also offer clues about the evolutionary origins of our own behavior.)

All this is not to say that every male chimp is a bully or that they are violent 24/7. They have different personalities—some shy, some sweet, some brutish. The big males, even Imoso, could be gentle and patient. They played with the youngsters, lightly wrestling and biting, and allowed their bodies to be used as jungle gyms as they tried to catch some shut-eye; they spent lots of time in their social groups, with the females, kids, and each other, traveling, relaxing, eating, and grooming, with little or no brutality. And although I saw very little aggression from the females, it does occur, sometimes with intensity.

And, of course, the same goes for adult males of the human community, who are capable of extreme acts of heroism, tenderness, and generosity, but also of violence and cruelty. I spent long hours every day as the lone woman in a group of local men, and I trusted them with my life. But during that time other men from the very same region of Africa were carrying out brutal acts against civilians.

The BBC World Service kept me company every night, and the lead story often featured the planet's alpha male, President Bill Clinton, and his affair with a young White House intern, Monica Lewinsky. As many men had done before him, and continued to do after, Clinton had risked everything for a few fleeting sexual encounters. Although this was a titillating distraction, I listened carefully for mention of the Congolese rebels, trying to glean any information on whether any were headed in the direction of my field site. A civil war was under way next door in the Congo, and the area was a hotbed of political violence. I learned of gruesome attacks, involving men using their machetes to attack villagers including children, cutting off hands, limbs, or heads, and raping the women. There were regular threats against Westerners, specifically of beheadings. I felt like a sitting duck, alone in my small bungalow at night, my own machete, tucked under my pillow, providing little comfort.

One horrific and widely publicized attack in March 1999 triggered the evacuation of most Westerners (including the Peace Corps) out of the region. Rwandan rebels had invaded a Ugandan national park, 250 miles to our south and also on the DRC border. The rebels killed four park employees and kidnapped fifteen tourists, who were marched into the mountains. The rebels slaughtered eight of them, from the UK, New Zealand, and the United States, with machetes and clubs. At least one woman showed evidence of a severe sexual assault.

I stayed on at my field site for another few months, but eventually, because of increased threats to Westerners and rebel movements in our area, the U.S. embassy ushered me out.

My experience in Uganda left me with the ambition of knowing more about how the shared biology of humans and nonhuman animals can help to explain why males and females are often so different. Really, I longed to understand men. Testosterone promised to be a key part of that explanation. So when my second application to Harvard was successful, and I started work on a PhD in biological anthropology, I learned everything I could about it.

INTRODUCING T

Testosterone is present in our blood in minute quantities. Both sexes produce it, but men have ten to twenty times as much as women. Despite its insubstantial physical presence, T has managed to achieve a substantial reputation, dwarfing that of any other corporeal chemical. After all, T is an "androgen," from the Greek "andro"—*man*—and "gen"—*generating*. If the Y chromosome is the essence of maleness, then T is the essence of masculinity, at least in the popular mind. Bill Clinton was assumed to have plenty of it, but with Donald Trump we got actual numbers.

Just before the 2016 presidential election, Trump appeared on Dr. Oz's national TV show to reveal the results of his latest physical. Oz read off the various numbers—weight, cholesterol, blood pressure, blood sugar. While the doctor seemed quite positive about what he described as "good numbers," only one number seemed to move the audience: 441 (nanograms per deciliter). Presumably, the audience's enthusiastic applause indicated that they saw Trump's T level as scientific proof that he possessed not just the spirit but also the constitution of a strong, masculine leader. And while the precise nature of the molecule itself is not especially enthralling to most people (its chemical formula is $C_{19}H_{28}O_2$), the same cannot be said for its ostensibly masculinizing powers—sometimes titillating, but sometimes toxic.

The writer Andrew Sullivan told readers of *New York Magazine* that he got "a real sense of what being a man is . . . [with] the rush of energy, strength, clarity, ambition, drive, impatience, and, above all, horniness" from his biweekly testosterone injections. A *Psychology Today* article suggests that "women are attracted to toxic masculine male phenotypes that correlate with testosterone . . . and who exhibit patterns of behavior that will allow them to ascend the social hierarchy and defend their positions from encroachers." According to the left-wing *Huffington Post*, Trump's presidency is "testosterone-fueled," making it "an extremely dangerous one" that could lead to war. According to the right-wing *American Spectator*, the problem is not too much T, but too little, among some prominent conservatives: "There is also a low-testosterone, dilettantish strain of conservatism that has overdeveloped in the 'mainstream' media . . . to create such sterile hybrids as Michael Gerson and George Will and David Brooks," who were, during Trump's first presidential campaign, "sipping tea" while Trump's base was "fighting a war." And in another piece from *Psychology Today*, the author describes the "testosterone curse" in which high T induces "a biological urge that sooner or later demands expression." According to him, while we can't forgive the sexual transgressions of Harvey Weinstein, Bill Cosby, and other male celebrities, we should understand that "men are just animals who, when under T's influence, have great difficulty perceiving females *other than* one-dimensionally, as objects for lascivious gratification."

So not only are powerful men suffering from the curse of hypermasculinity that leads them into war and rape, T is to blame, and we women can't help but love it! Apparently, too much T is toxic, too little is emasculating, and just the right amount leads to vigor and success.

Is any of this remotely accurate? Or is it just a popular myth—and perhaps one with a suspect sexist history? A proper answer to that question needs a whole book, and you are holding it.

There's no doubt testosterone is responsible for the human male's reproductive anatomy and physiology. As we'll shortly see, whether it's responsible for much more than that has been hotly disputed.

The consensus of experts is that testosterone's main job is to support the anatomy, physiology, *and* behavior that increases a male's reproductive output—at least in nonhuman animals. And men are no exception—T helps them reproduce, and directs energy to be used in ways that support competition for mates. How that works is the subject of the rest of this book.

SEX DIFFERENCES AND SEX HORMONES

Sex differences are simply differences between males and females—in humans, chimps, or other species—and noting a difference says nothing about its cause. Some differences are small, or inconsequential, at least for the purposes of this book: for instance, women are somewhat better than men at performing mathematical calculations like adding up numbers in a column. And women's names are usually different from men's names. Others are both large and meaningful. Men are much more likely than women to be sexually attracted to women, and they are far more physically aggressive than women in every pocket of the earth, at every age. For example, they are responsible for around 70 percent of all traffic fatalities and 98 percent of mass shootings in the United States, and worldwide commit over 95 percent of homicides and the overwhelming majority of violent acts of every kind, including sexual assault. One important point about sex differences, illustrated by these examples, is that almost any feature that differs between the sexes isn't exclusive to males or to females. Some men are called "Shirley"—and indeed the name was a man's name a few centuries ago. Women murder and sexually assault, they enjoy sex with other women, and many are slower and less accurate than most men when totaling up the household budget.

Let's look more closely at an obvious and uncontroversial sex difference: height. In the United States, the average height of women is less than the average height of men, by about five and a half inches. Following the pattern of many other sex differences, there is significant overlap: there are women who are taller than most men and men who are shorter than most women. If we picked hundreds of men and women at random and recorded their heights, the distribution of heights would look something like this:

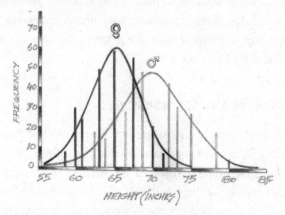

Sex differences in height: different average, different variation

The vertical axis (or "y-axis") represents the number of people from the sample who fall into each height category (in inches), which is indicated along the horizontal axis (or "x-axis"). The curves over each set of bars are just a clean way of approximating the (inevitably) messy data. (Only some bars are shown.) The dark bars represent women and the light bars men. Looking at the longest dark bar, it tells us that we found slightly fewer than sixty women who are sixty-five inches tall. We found more than twenty women who are seventy inches tall, and so on. The average height of women (at the top of the dark curve, around sixty-five inches) is clearly less than the average height of men (at the top of the light curve, around seventy inches), but there is a lot of overlap in height across the sexes.

The distribution of male heights is also *wider* than the distribution of female heights. The women cluster more tightly around their average than the men do around theirs. That is, there is more *variation* in male heights than female heights. This means that there are more men on the extremes of height, leading to more very short and very tall men, and fewer such women. More of the women are closer to the average female height than men are to the average male height.

A sex difference can be a difference in the average only (as we see in some tests of reading ability, in which females are higher), a difference in the variation only (as we see in IQ, in which there is greater variation in male scores), or a difference in both, as we see with height. The first two cases are illustrated in the graphs below.

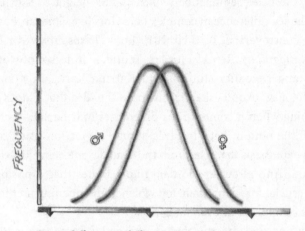

Group differences: different average, same variation

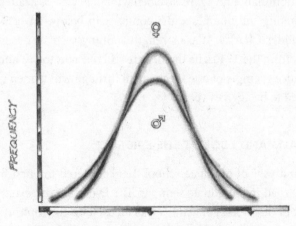

Group differences: same average, different variation

Sex differences are everywhere. Some are big, some are small, some are uninteresting, and some are striking and in need of explanation. One very large sex difference is the level of testosterone over a lifetime. What role—if any—does that sex difference play in all the others? One uncontroversial role of T is to increase the height of men relative to women. (Although, as we'll see in the next chapter, cutting off a boy's testicles before puberty *increases* his height.) Testosterone's role in the sex difference in complex behaviors like violence, however, is more controversial. In their 2019 book *Testosterone: An Unauthorized Biography*, Rebecca Jordan-Young, a professor of women's, gender, and sexuality studies, and Katrina Karkazis, a cultural anthropologist, express skepticism about the idea that T does much of anything when it comes to sex differences in behavior. According to them, the notion that "T drives human aggression" is a "zombie fact"—a hypothesis that rises from the dead despite being killed again and again. And elsewhere Jordan-Young writes that exposing this myth is crucial for "denaturalizing violence and opening up the remedies we can pursue or even imagine."

If T is not to blame, the obvious alternative hypothesis is that the greater level of male aggression is largely due to socialization. As the American Psychological Association puts it, "Primary gender role socialization aims to uphold patriarchal codes by requiring men to achieve dominant and aggressive behaviors." For a less academic way of illustrating the point, see the comic strip below, advertising the body builder Charles Atlas's strength training system. Even though the ad is from the 1940s, its themes are still relevant today, and it provides a nice example of one mechanism through which men could be socialized to be aggressive.

KEEP CALM AND LOOK AT THE EVIDENCE

In my first year of graduate school, I encountered my first bump in the PhD road in a graduate seminar, The Evolution of Sexual Behavior. One of our weekly meetings was focused on the topic of "forced copulation" in animals. One of the assigned readings was a research

Patriarchal codes

paper by the biologist Randy Thornhill, in which he set out a theory of the evolution of rape. Thornhill relied on the example of the male scorpion fly, which forcibly inseminates the female by holding her wings in an "abdominal clamp." According to the title of the paper, this is "rape": "Rape in *Panorpa* Scorpion Flies and a General Rape Hypothesis." From such behavior in the scorpion fly and other species, Thornhill speculated about the origins of rape in humans:

> Males should be most strongly selected to rape in species in which males provide resources important for female reproduction. . . . Rape is the only option for reproduction for a male without resources because he cannot deceive a female about his quality as a mate. . . . My hypothesis is that . . . in human evolutionary history, larger males were favoured because of the increased likelihood of successful rape if they failed to compete successfully for parental resources.

Whoa. He's saying that men evolved to be bigger than women so they can pin them down and rape them, just like the scorpion fly does, in the event that the men can't manage to impress the ladies with their ability to be a good provider.

That paper stirred an unpleasant feeling in my belly. When it came to my turn to contribute to the seminar discussion, I did my best to gather my thoughts. With my eyes wet and shining, I summed up my considered opinion to the rest of the group: "This guy is an asshole!" I still vividly remember how small, powerless, and angry I felt. Everyone seemed to be staring at me, waiting for an explanation. One other female student sat at the table with me, and I looked to her for validation—of course the men wouldn't get it. Nobody comforted me. Instead, the male professor calmly urged me to respond to the data and the argument. I thought, what's going on here? Is nobody else outraged? But he kept redirecting my attention to the evidence and reasoning in the paper. Eventually, I managed to overcome my revulsion and tried to assess the argument without letting my emotions get in the way.

It wasn't an easy process. My emotions didn't evaporate. And I'm still not thrilled with what strikes me as tone-deaf writing about a sensitive topic. But I learned that I could evaluate the evidence for an upsetting hypothesis on its merits; that by itself was empowering. (Incidentally, I briefly met Thornhill during my graduate studies, and he seemed like a perfectly nice guy.)

My students are often in the position that I was in on that day in class, encountering challenging ideas and research. Some react emotionally and reject them out of hand. That kind of response is easy to understand—emotional reactions, whether positive or negative, affect how animals, including humans, evaluate whatever it is they are encountering. If I see a big hairy spider in my bathtub, I can become aroused, and not in a good way—even if I know perfectly well that this particular species is harmless. That "spider stimulus" has caused unpleasant sensations in my body; therefore, the spider is bad. When we have a strong emotional or physical response to a stimulus—whether an arthropod, a person, an inanimate object, or a scientific hypothesis—we often irrationally project our response onto the stimulus itself. This can lead us to make poor decisions based on gut feelings rather than reasonable ones based on the proper evaluation of evidence. We can be driven to avoid accepting unpalatable conclusions.

The more I researched testosterone in humans and other animals, the more convinced I became that socialization is only part of the story. T, I have come to appreciate, plays a central role in human sex differences, and not just in physical traits. But as I was soon to discover, expressing that view carried perils of its own.

SUMMERS AND DAMORE

It was January 2005, and I'd recently finished my PhD in biological anthropology, transitioning from Harvard graduate student to Harvard lecturer. I had plenty of teaching under my belt, but always as a "Teaching Fellow," which is Harvardese for the assistant who meets weekly with small groups of students to discuss the material taught by the professor in lectures. I was thrilled to have the opportunity to create and teach my very own course and was busily preparing for the first class. The course material was based largely on my dissertation, which ended up being focused not on chimpanzees but on the role of testosterone in explaining sex differences in the way we think and learn, perceive the world, and solve problems. The class was to be a twelve-person seminar called Evolution of Human Sex Differences.

You may have heard of Lawrence Summers, who was then president of Harvard. Maybe you know his name because he was President Clinton's Treasury secretary, or because he has served as chief economist at the World Bank. But more likely, you might have the impression that he said something outrageous about how women are biologically unsuited for math and science.

That's not quite what happened. A few weeks before my course was due to start, Summers gave a talk at a small conference centered on how to get more women into STEM fields (science, technology, engineering, and mathematics). He offered several hypotheses that could explain why women are underrepresented in STEM. One invoked "different socialization and patterns of discrimination," which ruffled few feathers, if any. But another was that male aptitude was more variable (as male height is more variable), leading to more men than women at the very high (and very low) end:

> So my best guess, to provoke you, of what's behind all of this is that the largest phenomenon, by far, is the general clash between people's legitimate family desires and employers' current desire for high power and high intensity, that in the special case of science and engineering, there are issues of intrinsic aptitude, and particularly of the variability of aptitude, and that those considerations are reinforced by what are in fact lesser factors involving socialization and continuing discrimination. I would like nothing better than to be proved wrong, because I would like nothing better than for these problems to be addressable simply by everybody understanding what they are, and working very hard to address them.

Summers had hoped to stir discussion and debate with his comments. He at least succeeded in stirring the contents of the stomach of a well-known MIT biologist who was in attendance. She got up and left the room, and later told a reporter that if she had stayed she "would have either blacked out or thrown up." Accusations of sexism in the press soon followed. Donors stopped donating. Heated debates around campuses and water coolers ensued. After a vote of no confidence from the faculty, who considered his comments the last straw in a controversial presidency, Summers resigned under pressure.

It was no coincidence when over one hundred students showed up for my twelve-person seminar! And the controversy hasn't died down.

It was during the "Summers scandal" that I realized that I was on the wrong side of the divide. My embrace of evolution, testosterone, and sex differences seemed to make me morally suspect. I had taken for granted that to solve any problem (the underrepresentation of women in STEM, sexual assault, you name it), we must understand its roots, which can only happen in an atmosphere of free and open inquiry. That is, we need to be able to research, debate, and discuss all reasonable, thoughtful hypotheses, without shame or censure. That's what I had come to understand science and academia were all about. What's more, I expressed this sentiment to a reporter from the student paper—the *Harvard Crimson*—in response to his questions about President Summers's comments. I admit to being naive. I hadn't appreciated that some of my colleagues disagreed with me not only about the biological basis of sex differences but also about the kinds of questions that were suitable for discussion and investigation. One Harvard physics professor told the *New York Times* that it was "crazy to think that it's an innate difference—the difference in standard deviation. It's socialization. We've trained young women to be average. We've trained young men to be adventurous." He wasn't the only one to express such sentiments. It seemed that hypotheses like Summers's shouldn't be given airtime because such "dangerous ideas" might be discouraging to women and interfere with achieving gender equality.

At this time, most of the pushback that I felt came from male professors, who told me how things worked, and that any underrepresentation of women in STEM was due solely to discrimination and sexist socialization. But my research suggested otherwise. I was a new, untenured, female lecturer, and I soon began to harbor a sense of nervousness about how my opinions and abilities would be viewed by those on the top of the totem pole. I eventually gave up research and stuck to teaching, which I love. But in hindsight, I wonder if the environment around that time could have had something to do with my decision.

Fast-forward to 2017, and the annual ritual of updating the syllabus for my lecture course, Hormones and Behavior. I always begin the "Sex, Gender and Difference" unit by focusing on the role of T in fetal development, when it leads male development, in body and brain, to diverge from that in female fetuses. Once the students have the fundamentals under their belts, I use the Summers scandal as a way to introduce sex differences in behavior. What did he say? How was it covered by the press? Does the evidence support any of his claims? And should he have even suggested that biological differences might explain an undesirable situation for women? I thought about retiring my Summers routine because most of my students had never heard of him and in 2005 they had barely hit puberty. Luckily James Damore came to my rescue.

Damore would probably fit your image of a typical software engineer: male, a little nerdy. At the time he wrote his infamous internal memo—"Google's Ideological Echo Chamber"—in mid-2017, about 80 percent of the software engineers at Google were male. He thought that Google's efforts to achieve gender parity were misguided, resulting in a kind of reverse discrimination against men. In his three-thousand-word memo he wrote: "I'm simply stating that the distribution of preferences and abilities of men and women differ in part due to biological causes and that these differences may explain why we don't see equal representation of women in tech and leadership." And he implicated testosterone as the aspect of biology that mediated these differences.

The memo went viral, and soon Damore was the new Summers. One Google employee was quoted as saying that Damore's views were "violently offensive" and that she would never work with him again. Some cognitive scientists who reviewed his claims found them to be supported by the evidence, while others were more critical. But the relevant facts about sex differences had little impact on the emotionally charged response, nor did it prevent Google from firing Damore a couple of months later due to "advancing harmful gender stereotypes."

This was no doubt unfortunate for Damore, who subsequently sued Google, alleging "open hostility for conservative thought . . . paired with invidious discrimination on the basis of race and gender." But at least I had a more contemporary sex differences controversy for my new syllabus. That syllabus also contained many new articles on sex differences, reflecting the scientific progress that had been made since the Summers controversy. However, while the science

has moved forward, our ability to confront uncomfortable ideas suggested by science remains unchanged.

THE FEMINIST BACKLASH

It's all very well for me to sound these high-minded notes about keeping your emotions in check and dispassionately assessing scientific
hypotheses, but the fact is that women have good reason to be sus-
picious of "biological" explanations of sex differences. Scientists and
philosophers—mostly men—have a history of confidently expound-
ing on the alleged biological basis of women's inferiority. A prime
offender, I regret to say, is the greatest biologist of all time, Charles
Darwin. In his second book, *The Descent of Man, and Selection in
Relation to Sex*, published in 1871, Darwin provided evidence for
men's greater "mental power":

> The chief distinction in the intellectual powers of the two sexes is
> shewn by man's attaining to a higher eminence, in whatever he takes
> up, than can woman—whether requiring deep thought, reason, or
> imagination, or merely the use of the senses and hands. If two lists
> were made of the most eminent men and women in poetry, painting,
> sculpture, music (inclusive both of composition and performance),
> history, science, and philosophy, with half-a-dozen names under
> each subject, the two lists would not bear comparison . . . if men are
> capable of a decided pre-eminence over women in many subjects, the
> average of mental power in man must be above that of woman.

Men, he correctly observes, are greatly overrepresented on lists of
eminent thinkers and artists. But here it seems Darwin could not
shake off the Victorian cultural norms that characterized his time.
From our more enlightened perspective, we can create an obvious
alternative hypothesis: women are simply being held back by constraints imposed primarily by society rather than by their naturally
inferior mental capacity. Although a woman was the head of the British
Empire, women in Victorian Britain were generally not expected to

get an education. The University of London had first admitted women (a small group of nine) only a few years before *The Descent of Man*. And even then, they received only a "certificate of proficiency," not a proper degree. These days women have overtaken men in Darwin's own discipline and are awarded the majority (by a slim margin) of PhDs in the life sciences. Darwin, for all his greatness, got some important things wrong.

Lawrence Summers, you might have noticed, was arguing in a similar way, although his claim was that men have greater *variation* in "mental power" than women, not that the average is different. Summers is on firmer ground than Darwin, but since scientists are susceptible to prejudices and cultural biases like anyone else, caution is needed. People dismissed Summers's claims because his conclusion provoked strong negative emotions. But we should not ignore the possibility that Summers himself, or the scientists he cited, may have been too keen to find explanations that shore up the androcentric status quo. Bias goes both ways.

Bias affects everyone's thinking and work. The scientific explanation of sex differences can be subtly or not so subtly influenced by cultural norms to favor hypotheses that see sex differences as ordained by nature. For instance, women were kept out of professional sports until the early twentieth century on "scientific" grounds, as in this 1898 article in the German *Journal of Physical Education*: "Violent movements of the body can cause a shift in the position and a loosening of the uterus as well as prolapse and bleeding, with resulting sterility, thus defeating a woman's true purpose in life, i.e., the bringing forth of strong children." This is part of a long and broad history of using and distorting science in the service of dark ends. The eugenics movement in the United States is just one example: In 1931 twenty-nine states had laws that permitted forcible sterilization of those deemed genetically unfit. Former Harvard president Charles William Eliot pronounced eugenics laws essential to protect states from "moral degeneracy." Before the eugenics laws were abolished, close to seventy thousand people had been sterilized.

On the other hand, feminist critics of the science of sex differences can also be influenced by the fear that biology will be used to condemn women to domestic drudgery or otherwise reinforce the patriarchy. That concern may or may not be reasonable, but it is irrelevant to the truth of scientific hypotheses. And in the case of testosterone, it is hard to avoid the conclusion that these irrelevancies are motivating a lot of the critics.

If the problematic behaviors are socially created, presumably they can be socially destroyed. Alternatively, if the problematic behaviors are rooted in testosterone—and so are "natural"—what can we do? Short of castrating half the human race, we're screwed, right?

UNCOMFORTABLE IDEAS

Even if you hope that any disturbing conclusions about T's effects are not true, the point I want to emphasize is that this has nothing to do with whether they are true. In general, if you find a hypothesis distasteful, a red flag should immediately go up: there is a clear and present danger that you will discount the evidence that supports the hypothesis. That might seem obvious, but it is something that it took me a long time to learn and put into practice.

The idea that the gendered structure of human bodies, behavior, and institutions floats almost completely free from biology (and in particular from testosterone) is as popular now as ever. One leader of this movement is Cordelia Fine, a psychologist and the author of the 2017 book *Testosterone Rex: Myths of Sex, Science, and Society.* Fine thinks that the theory that testosterone plays a central role in masculine behavior is extinct, crushed to death by the weight of the evidence. Resurrecting this dinosaur, Fine says, is both fruitless and dangerous, and such thinking "squashes the hopes for sex equality." And if you believe that "biological sex is a fundamental, diverging force in human development," you're falling for an "overly familiar story" in which "differences between the sexes are shaped by past evolutionary pressures—women are more cautious and parenting focused, while men seek status to attract more mates."

Testosterone Rex was awarded the prestigious Royal Society Science

Book Prize, with one of the judges writing, "This book brilliantly explains how every baby, male or female, is born equipped to grow up into any sort of life." If we buy into that sexist story about evolution and hormones—the "Testosterone Rex" view—we will set severe limits on what we can accomplish. Removing those limits, Fine and others seem to believe, requires disproving the "ingrained myths" about biological differences between the sexes, especially those about testosterone.

The belief that biological explanations of sex differences inevitably lead to pessimism about progress, and a fatalistic acceptance of gendered social norms, is commonplace. As the neuroscientist Gina Rippon puts it in her 2019 book *The Gendered Brain*, "A belief in biology brings with it a particular mindset regarding the fixed and unchangeable nature of human activity, and overlooks the possibilities offered by our emerging understanding of the extent to which our flexible brain and its adjustable world are inextricably entangled."

A carefully curated reading list of books like *Testosterone: An Unauthorized Biography*, *Testosterone Rex*, and many popular magazine and newspaper articles would lead someone with little prior knowledge to wonder what all the fuss is about. If the science is so flawed, how did the myth of testosterone as the "male sex hormone" even arise? The journalist Angela Saini answers this question in her popular book *Inferior: How Science Got Women Wrong, and the New Research That's Rewriting the Story*. It is the clear and very real sexism in the history of science that has led us astray. In her view, only by exposing the bias and sexism in science can we see the real evidence. At the start of the book she asks: "Does the balance of sex hormones have an effect beyond the sexual organs and deeper into our minds and behavior, leading to pronounced differences between women and men?" And her answer is clear: "There are few psychological differences between the sexes, and the differences seen are heavily shaped by culture, not biology."

I agree with Saini that sexist assumptions can sometimes affect research. But I disagree on the answer to her question. Science shows that the answer is unequivocally "yes." In a number of important ways, testosterone pushes the psychology and behavior of the sexes apart.

In the following chapters, you'll learn how T influences our bodies, brains, and behavior in the service of reproduction. This isn't bad news, it's empowering information. Nothing we know about T or sex differences implies that we have to accept current levels of sexual assault, harassment, discrimination, or coercion. On the contrary, social progress depends on scientific progress. Understanding the forces that drive our priorities and behavior, and how genes, hormones, and environment interact, helps to equip us to combat the expression of the darker parts of our nature. There is no need to downplay the role of testosterone in our lives. Learning about how the world works and confronting the truth can sometimes be uncomfortable or disturbing. But I hope that it's mostly satisfying, empowering, and even fun, as it has been for me.

Internal Secretions

HANGING OUT OR IN?

Imagine a few different animals, all male: a frog hopping at the edge of a pond, an elephant grazing on the African savanna, a seagull circling overhead. And now imagine a man (naked, like the rest of the animals) walking his dog along the street. On which of the five animals can you see testicles? Since the image of frog and bird balls swinging in the breeze seems out of place, those two animals probably aren't on your list. What about the elephant? If you visualized dangling elephantine testicles then that's understandable, but wrong. He's one animal that would be particularly challenging to castrate. Like the frog and seagull, and most other vertebrates, the elephant's testicles are tucked away inside his body. The naked man and his dog? For both mammals, they are "descended." They hang down from the groin in the scrotum. Those precious, delicate organs, those sperm and testosterone factories, seem bizarrely vulnerable, suspended in thin-skinned sacs.

As a woman, I can only look on helplessly when an enjoyable game of touch football suddenly turns agonizing as one of the players throws himself to the ground, curled into the fetal position, writhing and groaning. Getting kicked, hit, or just banged in the balls appears to be excruciatingly painful. Next time that happens, it might provide some comfort to know that evolution put the pain there for a reason: if it hurts like hell, you will try harder to avoid a similar situation in the future. But just as you'd have to come up with a compelling explanation for choosing to keep all your cash in a paper bag on your front porch, evolution also needs to answer for letting that precious cargo hang out so vulnerably in the first place. Why aren't the testes always stashed away inside the body, like the heart and brain?

In all mammals, during embryonic development the testes start out in the abdomen, near the kidneys. And in most mammals, including humans, the fetal testes descend into the scrotum during the latter part of pregnancy, as a result of testosterone's actions. But in elephants and a few other mammals, such as the Cape golden mole (which looks like a mix between a small hedgehog and a hamster), seals, whales, and dolphins, the testes stay put in their original location inside the abdomen, just like the more sensible female ovaries. So what's going on?

Recent genetic discoveries suggest that the earliest mammals let them hang out. But as the mammalian evolutionary tree grew and branched, a few of those branches led to species with the genes for internal testes. Scientists aren't sure about exactly why those species took a different path, but external testes must do some good or else evolution would have eliminated them across the board.

One thing every guy knows about the scrotum is that it isn't simply an inert ball bag. When a man wades into cold water, he can feel the muscles in the upper scrotum (the cremaster muscles) contract to pull the testes closer to the warmth of the body—sometimes pulled in so tight that it hurts. And when he smushes them under his hot laptop, the muscles relax and loosen their grip in an attempt to get them to hang low, farther away from the body. We know that the scrotum acts as a climate control system, keeping the testes at a temperature that optimizes sperm production—about four degrees lower than the temperature inside the body. (If you want to ensure maximum sperm health, stay away from tighty whities and too many hours on the bike.)

Still, the mammals with internal testes also manage to keep them at the optimal temperature, by using different systems. The mystery of the species diversity in this testicular trait remains to be solved.

For someone who wants to understand hormones and how they relate to masculinity, dangling testicles are a stroke of luck. It's possible to remove the testes without killing their owner, and resulting changes in the animal can then be easily observed. Because of this relative ease of access, for more than two millennia, people have understood that the testes exert far-reaching influences on the appearance, behavior, and reproductive capacity of male animals. Modern behavioral endocrinology—the study of how hormones influence behavior—has its roots in this ancient knowledge of the power of the testes.

This chapter traces how this testicular knowledge gave rise to some truly bizarre (by present standards) social practices and set the stage for the experiments in the nineteenth and twentieth centuries that led to the discovery of testosterone. Hormones shape our brains and bodies to help us survive and reproduce, and in this chapter we'll look at the testes to begin exploring how T works its magic.

In the fourth century BC, Aristotle reflected on the changes wrought by castration, the removal of an animal's testes. In *History of Animals* he noted that the differences between "intact" and castrated animals were reminiscent of those between human males at different life stages (boyhood, manhood, or old age), and between breeding and nonbreeding animals, like loud and colorful male birds in the spring and their more subdued selves in the fall. Castrated animals provided evidence that the testes were responsible for the development and maintenance of certain masculine physical and behavioral features:

> Some animals change their form and character, not only at certain ages and at certain seasons, but in consequence of being castrated. . . . Birds are castrated at the rump at the part where the two sexes unite in copulation. If you burn this twice or thrice with hot irons, then, if the bird be full-grown, his crest grows sallow, he ceases to crow, and forgoes sexual passion; but if you cauterize the bird when young, none

of these male attributes or propensities will come to him as he grows up. The case is the same with men: if you mutilate them in boyhood, the later-growing hair never comes, and the voice never changes but remains high-pitched.... The congenital growth of hair never falls out, for a eunuch never goes bald.

A "eunuch," from two Greek words meaning "bed" and "to guard over," can refer to any man who has been castrated, or more specifically to those who also act as a servant or protector of a harem.

Whether to punish enemies or rapists, prevent the "mentally unfit" from having children, retain the high-pitched voice of a prepubertal boy, embody the identity of a woman, or to create a less lustful servant, castration has been common practice across cultures and ages.

THE CASTRATI

"Sistine Chapel Breaks 500-Year Gender Taboo to Welcome Soprano into the Choir." This was the headline from a 2017 story about the first woman to sing inside the Sistine Chapel (at least with the permission of the Vatican). The article quoted the famed Italian opera singer Cecilia Bartoli as saying she was in "seventh heaven" when given the opportunity to sing for an evening with fifty men and boys in the Sistine Chapel choir. What was all the fuss about?

The Holy See has never allowed women to sing in church. With Bartoli's gender-busting performance, a female voice had reverberated around the Sistine Chapel for the first time.

Business as usual resumed afterward: the Sistine Chapel Choir today remains exclusively male. How, then, without female singers to hit the high notes, can the choir fill its soprano parts? It relies on the voices of those males whose testes have not yet begun to produce testosterone (or the sperm it enables)—prepubescent boys. Soon their rising T will cause changes in their vocal tract, leading their voices to "break" and deepen. While there are exceptions, when those boys become men they can say good-bye to their angelic soprano voice. But there is a way to preserve that ability to hit the high notes, even giving it a boost when coupled with the large, strong lungs of grown men.

In the mid-sixteenth century, operas and choirs began to take advantage of the solution and filled their soprano spots with castrati: male singers who had their testes surgically removed before puberty had the chance to give them the sound of a man.

In spite of another Vatican decree prohibiting castration, by the mid-eighteenth century as many as four thousand Italian boys were subjected every year to the gruesome, dangerous, and painful procedure. (Anesthetics were still a hundred years away.) In spite of the risks, castration offered some families the chance to elevate their financial prospects, and they vied for their sons to obtain a coveted position as a castrato. A few castrati did achieve great fame and fortune, performing in opera houses all across Europe. But for the rest, prospects were bleak.

The lack of testicles and testosterone left castrati unable to marry and have a family, and changed their bodies in socially undesirable ways. If a castrato had his testicles removed before puberty got under way, then the changes were even more extreme. You may know that high T is what initiates the male pubertal height spurt, but you may be surprised to learn that it's also needed to end it. In puberty, rising T (or estrogen in girls) first acts to accelerate growth of the long bones in the arms and legs. But toward the end of puberty, peaking T (or estrogen) causes a cessation of bone growth. (More on how this works in chapter 5.) Without high levels of testosterone in puberty, boys miss out on the fast growth of the long bones that would usually occur during this time. But they more than make up for this with an extended period of "childhood" growth, continuing well past the point when adult height would have normally been reached, usually by age eighteen. They end up quite tall, with relatively long bones in the arms and legs, giving them a gangly appearance.

No matter when castration occurs, adult men without testosterone are fatter and weaker and have smoother skin than other men because normal male levels of T reduce fat and increase muscle, bone strength, and body hair (more on that in chapters 5 and 9). The ironic exception, noted by Aristotle, is that castrated men usually maintain an impressive head of hair into old age, since testosterone is responsible for most baldness. Most castrati were doomed to live their lives as social outcasts and were treated as freaks.

An eighteenth-century Italian boy wasn't usually consulted about the traumatic loss of his balls. After all, surely no sane man or boy would *choose* such agony!

THE EUNUCHS

Eunuchs were a feature of both ancient Greece and Rome. But there is probably no other culture with a longer, richer history of eunuchs than Imperial China.

Clear records date back to the Chou dynasty (about 1100 BC), but others suggest eunuchs served in ancient China as far back as the eighth century BC. Eunuchs continued to serve in the Forbidden City—the walled 180-acre opulent home of the emperor, his family, and their households, and the seat of the Chinese government—until the first part of the twentieth century, when the last emperor, Puyi, was driven out. Chinese dynasties relied on eunuchs to run many of the day-to-day operations of the government, to guard the virginity of the women who would bear heirs to the throne—the imperial harems—and to serve the imperial family. Because they had special access to high-ranking politicians and frequently had the ear of the rulers, the eunuchs possessed a great deal of knowledge about the inside workings of government. They were sources of gossip and advice and wielded significant political power.

Why were eunuchs trusted with these duties while men who possessed testicles were not? George Stent, an Englishman who spent many years living in China in the late nineteenth century, gave us the first comprehensive account of Chinese eunuch life. "There can be no other motive in using them," he wrote, "than the suspicion, distrust and jealousy of eastern potentates and others, of the fidelity of their wives and concubines; and the dread that if men were employed, licentiousness and profligacy would run riot in their harems."

The only men who would not pose a serious threat to the sexual and reproductive fidelity of the women in the imperial harem were those without testes, the sperm they produce, and a delivery system to transmit those sperm into a woman's body. And it didn't hurt that the lack of testosterone calmed or even eliminated sexual desire. The dynastic rulers needed strict assurance that only their true biological descendants would be the royal heirs. The solution was to forbid those outsiders with testes to come into contact with their precious virgins.

Some eunuchs were men who chose to trade their testicles, along with any hope of romantic love and a family, for the chance to escape poverty and be cared for in old age. But more commonly, young boys were coerced into surrendering their testicles and were often sold into servitude. The heyday of Chinese eunuchs coincided with the growing land shortages and famine that characterized the first part of the Qing dynasty (1644–1912), when the country's population began to surge and resources, including land for raising crops, were scarce. Citizens grew desperate as they struggled to feed their families. Just as it did for young male singers in Italy, castration offered Chinese men hope for a better life for themselves and their families. Serving the imperial family as a eunuch brought food and shelter and, for some, the ability to influence the course of Chinese politics.

The "last eunuch of China," Sun Yaoting, who died in 1992, described the anguish his parents felt about whether to have their son castrated. His mother was against it: "What does a small child know about it? Does he know it would make him a cripple all his life? Or that everyone would look down on him because he'd die sonless?" But his father was desperate: "Being a cripple is better than starving! Look at us. When will this life of poverty come to an end?"

The surgical procedure for producing a eunuch for employment in the Forbidden City didn't vary much across the dynasties and centuries. "Knifers," or *tao-tzu-chiang*, were trained men who would perform the ritualized surgery. For a fee, the knifers carried out the procedure just outside the gates of the Forbidden City, in a small building called the *Chang-tzu*—the shed. Inside, three assistants held the eunuch-to-be in a reclining position—one with his arms around the waist and holding down the arms, with two others spreading and pinning down the legs. The penis, testicles, and surrounding areas were bathed in hot-pepper water in preparation, ostensibly for anesthetic and disinfectant purposes. When the knifer was satisfied that all was well prepared, he brought his knife down to lop off both the penis and the testicles, which were preserved by the eunuch. The preserved organs, called the "*bao*," were said to be necessary for the return of a eunuch's masculinity in the afterlife and would be buried with him.

Recovery from such a grisly procedure, as one might imagine, was long and painful, and many new eunuchs died. And if the patient survived, he was likely to suffer serious complications. After castration, the knifer inserted a pewter needle into the urethra, now exposed at what had been the base of the penis, in order to maintain the opening to allow urine to pass out. Things went downhill from there:

> The patient is not allowed to drink anything for three days, during which time he often suffers great agony, not only from thirst, but from intense pain, and from the impossibility of relieving nature during that period. At the end of three days the bandage is taken off, the spigot is pulled out, and the sufferer obtains relief in the copious flow of urine which spurts out like a fountain. If this takes place satisfactorily, the patient is considered out of danger and congratulated on it; but if the unfortunate wretch cannot make water he is doomed to a death of agony, for the passages have become swollen and nothing can save him.

Sun Yaoting, the last eunuch, gave up not only his organs but also a conventional manhood to help his family. He had "dreamed of riches and honors." His life as a eunuch fulfilled some of his dreams, and brought pains and challenges he hadn't anticipated. "I have spent most of my life in the company of my eunuch friends. I have tasted both joys and sorrows."

With few exceptions, throughout the long reach of history, the ultimate purpose of removing the testicles of men (and other male animals) has been to deprive them of their some of their most quintessentially masculine features, such as physical strength, a deep voice, strong libido, or aggressive tendencies. Castration and its effects, while painful for the castrated, generated income and empowered animal breeders, politicians, and royalty. The idea that the source of masculinity was somehow located in the testicles has always fascinated philosophers and scientists, but they lacked knowledge of how this actually works until recently.

We now know that the testes are part of the endocrine system, the network of glands that regulates basic processes of animal life, like growth, metabolism, hunger and thirst, reproduction, circadian rhythms, and body temperature, in addition to related behaviors like eating, sleeping, fighting, parenting, and mating. We also know that the testicular masculinizing agent is the hormone testosterone. Mammals have at least nine endocrine glands, but the testicles are the only visible, readily accessible ones. Female animals lack testes, so have much lower levels of testosterone than males. If we want to understand what it means to be a male human, and the ways in which boys and men are different from girls and women, we need to understand T.

Hormone research didn't begin in earnest until the late nineteenth century, when scientists began to discern that the testes wrought their masculine effects via a blood-borne secretion, the precise nature of which would not be identified until the early twentieth century.

TRANSPLANTING TESTICLES AND THE ORIGINS OF ENDOCRINOLOGY

Humans aren't the only animals castrated in order to transform their appearance and behavior, and many of the undesirable side effects of castration in humans—fattiness and tenderness of the flesh—were sought after in other species. Cattle, pigs, sheep, and poultry are often subject to castration to limit breeding to the most desirable males and to increase the animals' agreeableness, in the farmyard and on the plate. Castrated cocks or "capons" have increased bone lengthening during puberty and grow to be significantly larger and fatter than normal chickens of either sex. Connoisseurs consider the meat of these animals a delicacy and describe it as buttery, tender, and juicy.

If you want to find out the mechanism by which the testicles work, chickens offer some advantages: they are cheap and plentiful and, once the procedure was mastered, not too difficult to castrate. The males and females are also spectacularly, unmistakably distinct. The rooster struts around with his shiny, colorful feathers, his head and neck adorned by a bright red comb and wattle, respectively; and his large body comes equipped with built-in weapons—sharp, bony leg spurs that he brings to battle against threats to his territory and the hens it contains. And of course, he amplifies his presence not only by his ornaments and weaponry but also with his powerful, cock-a-doodle-dooing voice. The hen is relatively understated in appearance and demeanor. Her color is duller, her size smaller, and her ornamentation less showy. And while she may get into the occasional skirmish and is capable of outright attack, she is peaceful by comparison.

In the early nineteenth century, the prevailing idea about how the testes conferred masculine traits was that they worked through the "sympathies," the system of nerves running through the body. The name arose from the observation that this system seemed to promote harmony among the various parts of an animal. (And the name has lasted—today, the "rest and digest" functions fall under the

Cock and hen

"parasympathetic" branch of the nervous system, and the "fight or flight" functions are part of the "sympathetic" branch.)

But Arnold Berthold (1803–1861), professor of medicine and curator of the zoological unit of the museum at the University of Göttingen, was unconvinced by the sympathies hypothesis. The logical alternative was that the testes somehow acted through the blood to affect the rest of the body and brain. Berthold set out to discover whether the testes would retain their widespread effects if they were transplanted. If the testicular transplant worked, then masculinization would take place in the absence of connections between the testes and the sympathies, i.e., nerves, and the alternative hypothesis would be vindicated.

Berthold experimented on cockerels—young male chickens. He castrated two of them, in the usual way, by making an incision in the belly, removing the testes, and sewing the birds back up. He knew how these birds would turn out—just like the castrati, their voices would fail to develop and they would look and behave more like hens. But then he castrated another two cockerels, and this time he put the testes back. Not where they belonged, though. He put one testis from each bird into the other castrated bird's abdomen. These two cockerels had testicles that were not their own and were in the wrong place. What were the chances that the misplaced, foreign testicles would rescue the cockerels from the feminization that occurs in castrated birds and other animals? Would they grow into adults with large and brightly colored feathers, wattles, and combs? Be loud, pugnacious, and lusty? Or would they become duller, smaller, quieter, calmer, and more reserved?

In his landmark article in 1849 Berthold described his observations: "So far as voice, sexual urge, belligerence, and growth of comb and wattles are concerned, such birds remained true cockerels." The transplantation worked and had retained the cockerels' masculine development! Upon killing and cutting them open, Berthold found that each testis had established a rich vascular connection with its host (specifically, each had attached to the colon) and had nearly doubled in size. The conclusion was inescapable:

Since, however, transplanted testes are no longer connected with their original innervation, and since . . . no specific secretory nerves are present, it follows that the results in question are determined by the productive functions of the testes, *i.e., by their action on the blood stream, and then by the corresponding reaction of the blood upon the entire organism.*

I still get goose bumps every time I read those last few words and think about their significance. Remember, the dominant idea at the time was that the nervous system somehow connected the testes to the rest of the body, and that's how they managed to confer such large, systemic changes in behavior and appearance. Berthold found that the testes exerted their physical and behavioral effects through the blood, without any nervous connection. This was revolutionary. The testes must have passed something into the circulatory system that prevented the feminization that occurred in the first set of castrated chicks.

Berthold was not the first to postulate that the testes acted through the blood, but he was the first to publish experimental data in support of the hypothesis. His discovery that testes acted on the blood, ultimately affecting behavior, set in motion the development of the field of behavioral endocrinology. But it would be another ten years before Darwin sketched out the first clear scientific framework for understanding why such differences existed in the first place: sexual selection. Stay tuned for more (in chapter 6), but briefly, this is the idea that evolution equips males to compete for mates, thus explaining why they are frequently larger, brighter, or more pugnacious than females.

And a century later, testosterone would finally be isolated and identified, solidifying the foundation upon which a robust body of knowledge about the biology of masculinity would be built.

But it was still early days. Berthold's discoveries suggested to some that the testes contained some magical masculinizing potion, and perhaps manliness could be medicalized. It didn't take long for that idea to catch on.

THE FOUNTAIN OF YOUTH

Reduced vigor, along with wrinkles and fading faculties, are consequences of aging. We may try (and succeed) to mitigate some of the
effects by maintaining a healthy diet and ensuring we get enough
exercise, and accept the inevitable slow decline. But the antiaging
industry would like us to consider other options. We can avail ourselves of products and services like Botox, expensive eye creams, or
various kinds of supplements that claim to increase energy, strength,
even sexual performance. You might be surprised to learn that this
industry has deep historical roots—in the testicles.

The esteemed German anatomist and zoologist Charles-Édouard
Brown-Séquard (1817–1894) had been a prolific researcher with over
five hundred scientific papers to his name. But the loss of energy and
scientific productivity that he experienced in his seventies was intol-
erable. Having devoted most of his career to the study of the nervous
system, he turned his focus to "internal secretions"—a concept that
he introduced during a talk at the Society of Biology in Paris in 1891.
He postulated that the cause of disease could be discovered by treating
animals (including himself) with the extracts of various tissues; if the
treatment worked, then the cause was due to inadequate secretions
from a particular tissue or gland.

Brown-Séquard had become obsessed with the potential powers
of "testicular secretions" in particular. The loss of semen through sex
or masturbation resulted in lethargy, it was thought at the time, so
testicular secretions might restore it. As he put it in an 1889 article,
"Note on the Effects Produced on Man by Subcutaneous Injections of
a Liquid Obtained from the Testicles of Animals":

> It is known that well organised men, especially from twenty to thirty-
> five years of age, who remain absolutely free from sexual intercourse
> or any other causes of expenditure of seminal fluid, are in a state
> of excitement, giving them a great, although abnormal, physical and
> mental activity. These two series of facts contribute to show what

great . . . power is possessed by some substance or substances which our blood owes to the testicles.

Never one to shy away from self-experimentation, Brown-Séquard had begun to inject himself with the extracts of crushed testicles, including blood and semen, harvested from guinea pigs and dogs. In that same Paris talk, Brown-Séquard enthusiastically relayed the astonishing effects. The arc of his urine—clearly the most important measure of manliness!—had become longer, and his mental clarity and focus had increased, as had his handgrip and other measures of physical strength and stamina. His lively and detailed descriptions of the potency of these internal secretions, which lacked the scientific rigor and care of his previous work, were met with skepticism by many of his scientific colleagues. But those eager for fast cash or easy fixes were far less critical.

A booming industry of "organotherapy" for the normal symptoms of aging and various other ailments began. The "Brown-Séquard Elixir" was prescribed by thousands of physicians and even more charlatans. In addition to organ extracts, tissue grafts and cellular transplants also became part of the—then mostly ineffective— medical armory. Testes from accident victims, executed prisoners, rams, and goats were transplanted into those willing to pay for the hope of rejuvenation. This quackery continues today. If you feel listless, your libido is falling, your erection not quite up to snuff, your muscle mass down, just click a few links on the internet and delivered to your front door will be a version of Brown-Séquard's elixir.

The restorative powers of organ extracts that Brown-Séquard reported so convincingly were almost certainly due to the placebo effect (knowing that might also be true of my eye cream never stops me from using it!). Although he jump-started a dubious medical industry, Brown-Séquard did leave a positive legacy. As he presciently remarked in the British medical journal the Lancet, "The results show that this important subject should be further investigated experimentally." Even though his results were not real, he helped the scientific investigation of hormones to take off.

THE "CRUCIAL EXPERIMENT" AND BULLS' BALLS

In the wake of a growing fascination with "internal secretions," Ernest Starling, a respected British physiologist, along with his brother-in-law, William Bayliss, sought to determine the mechanism through which the pancreas secreted sodium bicarbonate, which counteracts the acidic digestive juices in the stomach and small intestine. (Without sodium bicarbonate, the active ingredient in baking soda, you would suffer intestinal burns. You may have taken baking soda yourself to ease an upset stomach.) Bicarbonate is not itself a hormone, but the pancreas has to know when to release it. Bayliss understood that an internal signal like "Too much acid, please help neutralize it!" must be sent from the digestive organs and received by the pancreas. But, echoing Berthold's research on the means by which the testes communicated with the rest of the body, Bayliss aimed to determine whether the pancreas communicated with the digestive organs through the blood, or alternatively, through the nervous system. The latter explanation was generally accepted and favored by the world's most influential physiologist at the time, a Russian called Ivan Pavlov. (You may remember Pavlov from his famous experiments on dogs.)

Dogs were also Bayliss and Starling's experimental subjects, although their dogs had a more miserable fate. The subject of Bayliss and Starling's "crucial experiment," as they called it, was anesthetized with its digestive system surgically exposed. They discovered that when acid was introduced into the small intestine, it produced a "well marked secretion." Bayliss and Starling, as they reported in a classic 1902 paper, were able to determine that this secretion affected the pancreas through the blood:

> We soon found, however, that we were dealing with an entirely different order of phenomena, and that the secretion of the pancreas is normally called into play not by nervous channels at all, but by a chemical substance which is formed in the mucous membrane of the upper parts of the small intestine under the influence of acid, and is carried thence by the blood-stream to the gland-cells of the pancreas.

Bayliss and Starling called that chemical substance, the first hormone to be isolated, "secretin." Pavlov, who had hypothesized that the intestines communicated with the pancreas via the nervous system, attempted to replicate Bayliss's experiment, apparently hoping that it would fail, so that his theory would be shown to be correct. But Pavlov didn't let his pride get in the way of science: seeing firsthand that the pancreatic response to stomach acid relied on chemical signals in the blood and not the nervous system, he exclaimed: "Of course they are right. It is clear that we didn't take out an exclusive patent for the discovery of the truth." Two years later Pavlov won the Nobel Prize, "in recognition of his work on the physiology of digestion, through which knowledge on vital aspects of the subject has been transformed and enlarged."

During a lecture at the Royal College of Physicians in 1905, Starling described his discovery of secretin and what he had learned about the differences between neural and chemical control of bodily functions. He also coined a name for messengers like secretin, borrowing from Greek:

These chemical messengers, however, or hormones (from *ormao*, "I excite or arouse,") as we might call them, have to be carried from the organ where they are produced to the organ which they affect by means of the blood stream and the continually recurring physiological needs of the organism must determine their repeated production and circulation throughout the body.

The discovery of secretin opened up a new window into the basic physiological functions of organisms: chemicals, produced by special glands, traveled through the blood and affected distant tissues, regulating and coordinating bodily functions. But that was just the beginning, and discoveries in the field moved rapidly. In a short period beginning in 1929, the three estrogens (including estradiol, commonly called "estrogen," along with the far less abundant estrogens, estriol and estrone) were discovered. Shortly after that came the discovery of testosterone.

At the end of the nineteenth century and into the early twentieth, extracts from the testes offered hope for the tired, elderly, effete, and impotent. But the active agent had not yet been identified, let alone synthesized in large quantities. As Brown-Séquard's Elixir and its competitors were gradually discredited, other scientists were hard at work to identify the masculinizing secretion. The basic scientific knowledge, laboratory techniques, and financial incentives were all in place. It was only a matter of time.

Gone were the days of crushing pig and monkey gonads. The new technique—harvesting organs from slaughterhouse animals—proved more productive. One physiologist at the University of Amsterdam, Ernst Laqueur, even cofounded a company called Organon (still in business today, as a subsidiary of the Merck pharmaceutical company) in close proximity to several slaughterhouses, providing him with easy access to the testicles of slaughtered bulls. In 1935, he collected 100 kg (about the weight of a baby elephant) of bull testes. From these, he harvested a minute 10 mg (less than the weight of a grain of rice) of a chemical that he injected into a castrated rooster to determine the extent to which it could regenerate his comb. (This had become a standard test to determine the extent to which a particular substance could induce masculine traits.) The chemical regenerated the cock's comb, replicating the effects of testes transplantation. Laqueur dubbed it "testosterone." By 1935, three research teams led by Adolf Butenandt, Károly Gyula David, and Leopold Ružička, who were individually sponsored by three different European pharmaceutical companies, published papers on the synthesis of testosterone almost simultaneously. Butenandt was awarded the Nobel Prize in Chemistry in 1939 for his work.

Today, around seventy-five hormones are known. The endocrine glands (depending on how you define them) at least include the hypothalamus, pituitary, thyroid, parathyroid, adrenal, pineal, ovaries, testes, and pancreas. But in addition, as Bayliss and Starling showed, not only the endocrine glands secrete hormones. The intestinal and fat cells, for example, are active hormone producers. The

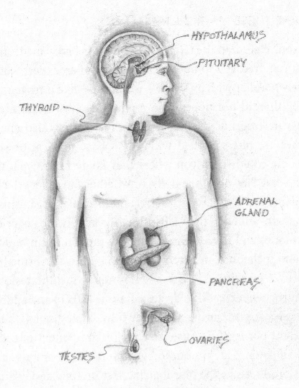

HYPOTHALAMUS

PITUITARY

THYROID

ADRENAL GLAND

PANCREAS

OVARIES

TESTES

Major endocrine glands

list goes on to include the liver, heart, kidneys, skin, and importantly, the brain—they all secrete and respond to hormones. Even your gut bacteria secrete hormones, some at very high levels, and many with functions that have yet to be identified. Every new discovery about the endocrine system surprises us with the inventiveness of natural selection in keeping us alive, healthy, and reproducing, and suggests ways we can build on that inventiveness to treat the sick or improve people's lives. Finding the truth amid the hype around T will be impossible without a basic understanding of hormones in general, so let's take a look at this vitally important class of chemical communicators, of which T is a member.

WHAT ARE THESE MAGICAL MESSENGERS?

Unless you've studied hormones in school, or had a medical condition that required you to become somewhat of an expert, you might be vaguely familiar with only a few hormones—like estrogen, insulin, or maybe thyroid hormones. You might have heard that melatonin can help you sleep, but you may not have realized it is also a hormone. Hormones are often imbued with all sorts of magical powers, but most people couldn't tell you where they come from, what they are made of, what they do in the body, or whether they affect the brain.

All plants and animals—in fact, all multicellular organisms—have hormones. In animals, they fall primarily into two categories: *protein* hormones and *steroid* hormones. The protein hormones include insulin and melatonin and are made from amino acids (the building blocks of all protein). The steroid hormones include testosterone, other androgens such as dihydrotestosterone (DHT) and androstenedione, and estrogen; they are all made from cholesterol. (Cholesterol is important not least because it's a major component of cell membranes.) Hormones are produced by various glands and tissues: melatonin is produced by the pineal gland; testosterone and estrogen are

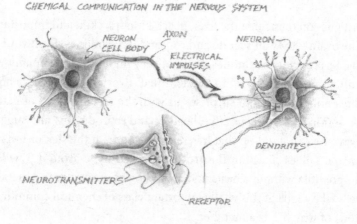

Neural communication

produced by the testes and ovaries (and other tissues); and insulin is produced by the pancreas. All of our hormones circulate in our blood and can be thought of as carrying information to various parts of the body. Since hormones go anywhere the blood goes, they circulate pretty much everywhere.

In any complex system, whether it is an organism, a family, a widget factory, or a university, information needs to be communicated from some parts to others to keep the whole system functioning properly. In the complex system that is our body, this communication is carried out by "chemical messengers." Animals have two main types of such messengers: neurotransmitters—facilitating communication in the nervous system (the brain and spinal cord)—and hormones.

While neurotransmitters convey information via electrical impulses in a point-to-point fashion via neurons, like trains traveling on a set of branching tracks, hormones send their chemical messages out far and wide, to any cell that is "listening." My favorite radio station in Boston is 90.9 WBUR, and I can only receive their content if my tuner is set to the right frequency to pick up the signal sent out by WBUR's radio tower. Hormones are broadcast and received like that tower and my tuner. They are sent out by the hormone-producing glands (and other hormone-producing cells) into general circulation (the blood system), but their signals are only detected by cells that have

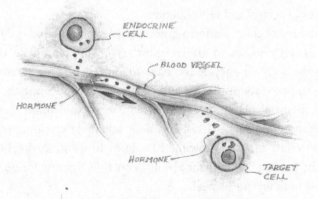

Hormonal communication

specific *receptors* for them. The network of endocrine glands and cells makes up the *endocrine system*. Cells that can respond to a given hormone are the hormone's "target cells." The receptors for the protein hormones are embedded in the outer membranes of target cells, and receptors for the steroid hormones, including testosterone, are located on the inside of target cells. (There are some exceptions to this, which we'll get to later.) A cell that doesn't have receptors for testosterone is just like a radio that is tuned to WGBH (another of my local stations). The radio may encounter WBUR's signal but can't pick it up; likewise for the cell that encounters testosterone but isn't "tuned" to it.

As you read this now, neurotransmitters are helping you decode and respond to the visual stimuli that are these words and translate your desire to keep reading into the action of turning the page. If you're hungry, hormones have sent signals about your energy status from your gut to your brain, where they alter how you feel, causing you to struggle to stay focused on testosterone while you crave a little snack. That motivation to get food was triggered by the interaction between the endocrine and the nervous systems (the interface between the two systems is called the "neuroendocrine system"). This is how hormones influence your brain and behavior. They facilitate communication between your body and brain (and vice versa), to coordinate your desires and actions with your physical needs. Hormones and neurotransmitters are there because the blind forces of evolution put them there, which is to say that they are ultimately in service of survival and reproduction.

This kind of coordination is crucial to meeting your body's demand for the energy needed to survive, grow, repair, and reproduce. Energy is like the body's money. A company might rely on email communication to know about income and expenditures, informing the company's financial planning. But how do a body and brain exchange information and make decisions about saving or spending energy? When to start and stop growing? Should we invest in staying healthy, or is there enough energy for playing, courting, fighting for mates, or making milk for our babies?

If you're on the way out for a run, you might prepare by taking a few bites of an energy bar. You're a mile into your run, and the cells in your working muscles are very interested to know if energy is readily available. Your body has plenty of options about where to get the energy your cells need, but it's easier and energetically cheaper to take some glucose directly from the blood without having to create it from other sources like fat. Just in time, the glucose from your energy bar is hitting your bloodstream. That glucose interacts with special glucose-sensing cells in your pancreas, signaling its presence in the blood and its availability for the rest of the cells in the body, especially the ones that are working hard. In response, your pancreas releases insulin, which also enters your blood, circulating far and wide. It shouts to cells, "Hey! Open up! Glucose is at your door!" Almost all of your cells are targets for insulin, since they all need glucose for energy, but those in working muscles are especially eager to respond to insulin's signal. Muscle cells (and other needy cells) respond by opening their doors to let glucose in. When insulin binds to its receptors on your muscle cells, it causes a special portal to open in the cell membranes, through which glucose can flow. Since you're out for a run and working hard, your cells need energy right away. So the glucose that enters your hungry cells is immediately used to produce ATP, which is a molecule that they can use for fuel. Thus, insulin sends the information about the availability of glucose to the whole body (the brain uses different mechanisms to get glucose into its cells). Working cells responded, and you now have the energy to finish your run.

But a little while after you get home and crash on the couch, you feel the urge to get up and rifle through your snack drawer for something satisfying. And there's a good reason: insulin spiked up in response to the carbs you ate earlier, and it did its job well. It caused your cells to take lots of glucose out of your blood. Now your blood-glucose levels are a bit low, which means insulin is also low. Your brain senses the low insulin, and its appetite centers take action when they get insulin's "low blood glucose!" message. They signal other parts of the brain, via neurotransmitters, to link your low glucose situation to a behavior. You relieve your nagging discomfort by moving your body up and off the couch. You may grab an apple out of your fruit bowl. But if you're like me, you end up with a handful of pretzels or potato chips.

As your gut breaks down the food into macromolecules that can be absorbed from your small intestine into your bloodstream, your blood sugar goes up, and so does insulin. It moves that glucose into cells, and the high insulin sends that information to your brain, telling it that you've got plenty of energy. You're no longer hungry. The hormone has interfaced with your brain, coordinating your behavior with the needs of your body, which, for now, are largely met.

Testosterone (along with the other sex hormones) is like insulin: it acts in the body, and it also sends information to your brain. Like insulin, testosterone coordinates physical and behavioral processes, but rather than regulating blood sugar, its focus is on developing and supporting reproduction.

Of course reproduction requires energy, so the functioning of your reproductive system depends on how much energy is available (and hormones like insulin and testosterone always work in concert with other hormones and other chemicals). And that relationship, between energy and reproduction, is much more constrained for women than for men. For men, successful reproduction has less to do with energy-sucking reproductive physiology—growing and feeding a baby with one's own body—and more to do with finding, competing for, and attracting mates. If T is relatively high (in the typical male range), it works in a man's body to promote muscle growth and sperm production, and it also tells the brain what the body is up to. High insulin shouts, "Lots of energy down here, use it!" High T shouts, "Lots of sperm ready to go!" Testosterone helps males do what they need to reproduce—as castration throughout the ages has shown.

In the absence of modern pharmaceutical treatments, typical male behavior and appearance depend on having testicles. As we've seen, humans have long observed that without testes, male animals fail to develop (or lose) masculine physical traits and processes like the production of sperm, and (depending on the animal) bright feathers, large, pointy antlers, a long larynx, or big upper-body muscles. And the testes' influence extends to masculine behavior, like getting an erection, using physical aggression, and being motivated to compete for female sexual attention.

Starting in the late nineteenth century, research and industry focused on testicular secretions boomed. Scientists could examine the effects of hormones on animal physiology, anatomy, and behavior without resorting to the laborious, minimally effective, and relatively cruel methods of transplanting entire organs or tissue grafts. By the early twentieth century, estrogen, progesterone, and testosterone were being synthesized and sold for use in pharmacology (for quack and serious medical purposes) and for research.

In addition to the lifesaving discoveries and the synthesis of new drugs such as insulin—first used as a treatment for type 1 diabetes in 1921—early partnerships among clinicians, pharmaceutical companies, and researchers made the field of endocrinology flourish. This symbiotic relationship continues today, leading to an ever-greater understanding of the ways in which hormones mold the body and the brain.

Prior to the discovery of testosterone, all evidence about the hormone's actions on animals came from castration—removal of the source of testosterone *after* birth. But I haven't yet mentioned that the normal development of most male animals requires testosterone exposure *before, or shortly after* birth. So what happens when a male fetus does not experience T's actions in the womb or at any time thereafter? That is a question for the next chapter.

3

JUST ADD T: MAKING BOYS

JENNY

Jenny appeared at my office door right on time. Many students wear a daily uniform of jeans or sweats and T-shirts, but Jenny looked put together as usual, in a red dress that fell to just above her knees and black patent leather flats. Her highlighted light-brown hair flowed neatly down past her shoulders and set off the white of her pearl stud earrings.

Classes were over, and final exams were looming. Unlike most students who visit my hours during what we call "study period"—the time between the end of classes and final exams—Jenny didn't come with questions about what would be on the exam for my Hormones and Behavior class or how to study for it. Instead, she smiled warmly and said that she wanted to "talk about the course." The proposed discussion topic piqued my interest, particularly in light of the fact that she had never before dropped by my office hours. But I knew Jenny's face—her animated blue eyes, high cheekbones, and unblemished, pale skin—mostly because it was one among the sea of faces that drew my gaze as I taught. Her alert expressions helped to keep me feeling that I was getting through. She always sat up straight, focused on me or her notes, and she would nod along to what I thought were the more interesting points.

On a couple of occasions she'd also been part of the posse of students that gathered around the lectern after class, notebooks at the ready. The questions tended to come in two varieties: those designed to solidify understanding of new terms or concepts—"Which cell types produce Müllerian inhibiting factor again?"—and those that reflected deeper engagement with the material—"Were the researchers who recorded sex differences in activity level aware of the sex of the babies?" Jenny's questions were of the second type.

I welcomed her into my office and we sat across from each other at a small round table. We got to chatting about her interests, her family, and her time at Harvard. It all seemed consistent with my expectations. She was enjoying her classes, sang soprano in a campus choir, and was active in her sorority (technically, Harvard doesn't have sororities, but her women's group was very much like one). Like most of my students, she hoped to go to medical school after college. Given her smarts, work ethic, and compassionate nature, I was sure she would be a success. She came from a close-knit southern family. She seemed happy, well adjusted.

Then Jenny said she wanted to tell me about how she had become interested in hormones. Over the years, lots of students have shared stories about what had brought them to my class: they or a relative have an endocrine disorder like diabetes or hypothyroidism, they're transgender and taking cross-sex hormones, or they're interested in bodybuilding. Often these students are extremely knowledgeable about the endocrine system despite not having taken any previous coursework.

Jenny began to describe how, during her teenage years—the time when most of us just want to fit in—she learned that she was different from her peers and friends. She relayed her personal story in a straightforward, cool manner. I listened, stunned, and struggled not to tear up.

When her girlfriends started getting their periods around age twelve, Jenny assumed that she would soon be passing through this rite of womanhood herself. She had all the usual signs of puberty that her friends were showing—her breasts had grown, her hips had widened, she had put on fat in the typical places. But she turned fourteen, then fifteen, and still no period. Even though she seemed to be a healthy teenage girl, not getting her period by age sixteen could be a sign of an underlying health problem. So Jenny's mother took her to a gynecologist to get checked out. It turned out that things were more complicated than they had assumed, and Jenny was not just a late bloomer. She was referred to a specialist for further tests, and blood was drawn, an ultrasound performed, her body probed. A team of doctors called Jenny and her parents in to discuss the results.

Jenny learned that day that her health was fine, which was a relief to everyone. But she also learned that her body was different—she had a difference (sometimes called "disorder") of sex development (DSD). Jenny had a rare DSD known as complete androgen insensitivity syndrome (CAIS), which occurs in about two out of every hundred thousand people.

The medical label CAIS does a good job of describing the primary issue. Instead of XX sex chromosomes inside her cells, typical of female mammals, Jenny's cells contained an X and a Y, typical of male mammals. Her Y chromosome led to the development of testicles rather than ovaries, but those testicles remained in her abdomen, undescended. They produced lots of testosterone but her body couldn't respond to it, or to androgens in general. And because her body couldn't respond to T, her testes could not make sperm. That absence of testosterone action allowed her female traits to develop in utero. Jenny had a normal-appearing vagina, but it did not connect to a uterus. As a result, she also had to confront the reality that she could never become pregnant.

That life-changing conversation was disorienting and frustrating for Jenny and her family. But as she later came to appreciate, she was fortunate. Since learning of her condition, Jenny had connected with other CAIS women who had been through similar experiences, and they shared their stories with her, sometimes becoming close friends.

Many of them told Jenny that they had been lied to, given poor medical advice, subjected to unnecessary surgery, or made to feel ashamed of their bodies. Jenny knew how lucky she was to have a supportive family and a team of doctors who were sensitive and progressive in their approach to seeing this as a lifelong condition and who recognized the harms and benefits of each potential decision.

CAIS is often described as an "intersex" condition, in which a person's external reproductive anatomy is not what we would expect given their gonads or sex chromosomes. Such conditions fall under the broader DSD category (which you will learn more about soon). Understandably, some people have strong feelings about the appropriate terms. Jenny prefers "variation" for CAIS, so that's what I'll use.

With XY chromosomes and testosterone-producing testes, but without T's actions, a person will have female primary and secondary sex characteristics. ("Primary sex characteristics" are the internal and external genitalia and are present at birth; "secondary sex characteristics" are those features like breasts in females and facial hair in males that appear at puberty.) The appearance and behavior of people with CAIS are unexceptional; that is, they look and act like typical women or girls. In many cases, it's only the lack of a first menstruation that clues CAIS girls into the fact that they have a body that's different from that of their peers.

I had been teaching about the science of CAIS for years, relying on it as an example that demonstrates the astounding power of testosterone, but I'd never known someone who was living with it. Here was someone whose body contained what I had thought of as keys to masculinity—testicles and testosterone. But Jenny struck me as ultrafeminine. Despite my knowledge of her variation, I struggled to come to terms with what still seemed like a mismatch. There are many combinations of sex chromosomes, gonads, and sex hormone levels that contribute to our looking and feeling masculine, feminine, or something in between. Masculinity and femininity don't always come in the packages we expect.

Her story pulled at my emotions, and even though she seemed confident and at peace with herself, my sympathy for her and the challenges I imagined she had faced surged. But at the same time, I couldn't help thinking, "Shit! Is she here to teach me a lesson about how I should handle this material in the classroom? Have I screwed up royally and offended or hurt her with some insensitive comment?" If I had, she chose not to mention it. Instead, she asked if I would consider working with her the following semester, engaging in an "independent study," in which the two of us would learn as much as we could about CAIS. Of course, I said yes.

BAKING BOYS AND GIRLS

You may have thought that possessing XY sex chromosomes in (nearly) every cell is incompatible with the natural development of a vagina, not to mention breasts, wide hips, a high voice, and soft, smooth skin. Jenny shows us that it isn't. The sex chromosomes themselves don't stimulate (or inhibit) those feminine traits. That's the job of the sex hormones, and sometimes they don't align as expected with the sex chromosomes.

The reason that boys develop a penis and girls do not, or that men have lots of coarse facial hair and women develop breasts, is not that the genes for these traits are possessed solely by one sex or the other. Females don't have exclusive ownership of genes that result in the development of milk-producing breasts or wide hips, and males do not have sole rights to the genes for a deep voice or facial hair. Both sexes come genetically equipped to express almost all the traits typical of either sex. It's just a matter of which genes are active, at what levels, in which bodies. This fact will not come as a shock to women who have doled out thousands of dollars ridding their face of dark hairs, or to men who struggle with gynecomastia, disparagingly known as "man-boobs." This is why transgender people can have great success in taking on the physical traits of the other sex. Humans have some genes that differ by sex, because males typically have a Y chromosome and females don't. But the number of genes on the Y is minute—about seventy—relative to the twenty thousand to twenty-five thousand genes that populate the other twenty-two pairs of chromosomes. But don't underestimate the power of the diminutive

Y, which can pack quite a punch. One of its genes makes all the difference in the world.

To understand how a male and a female, who have nearly identical genes, can have vastly different body plans, imagine getting ready to bake some cookies. Your kitchen is well stocked with all the ingredients you need to make a variety of cookies—butter, brown and white sugar, baking soda and powder, flour, chocolate chips, oatmeal, nuts, etc. You can make any of hundreds of cookie recipes, but your friend has put in a request for chocolate chip.

You turn to the appropriate page, read the recipe, gather the ingredients, mix and bake them, and you and your friend enjoy the warm, chewy, chocolaty treats.

When Jenny was a ball of dividing cells in the womb, and then a mass of growing and differentiating tissues, her versatile stem cells could become a number of different cell types. Just as I can choose a recipe to follow to make a particular type of cookie, the cells in growing embryos choose to "read" specific genes to make proteins, forming various cell types such as muscle cells, red blood cells, nerve cells, and more.

The forty-six chromosomes in each of our cells contain the whole human "genome"—all of our DNA (deoxyribonucleic acid). DNA is a molecule shaped like two long springs pushed sideways into each other. End to end, there's about six feet of it in each cell, and all the DNA from all of your cells could stretch to the sun and back two hundred times. Your genes reside in your DNA and are strings of "chemical letters" (or "bases") from a very small alphabet—A, C, T, and G—that provide the instructions for making proteins.

So each gene is a kind of recipe for a protein. (In the jargon, the gene "codes for" the protein.) The gene lists the necessary ingredients and specifies the order in which they should be put together. But instead of combining butter, sugar, and flour, proteins are made by stringing together chemicals called amino acids. In humans, there are twenty-one types of amino acids to choose from. You may have heard of a few of them: the amino acid phenylalanine is used to make the artificial sweetener aspartame, and the amino acid tryptophan found

Baking as gene expression

in turkey is said to make us sleepy after the Thanksgiving meal. (That last one's a myth, by the way!)

Take the gene for the hormone insulin, which calls for fifty-one amino acids to be strung together. Since there are only twenty-one types, some amino acids will be used more than once, just as the ingredients for one cookie might include ten chocolate chips and four walnuts. Cookie recipes need to be *read*, and then the ingredients need to be *mixed and baked*. In the case of genes, they are *transcribed* and then *translated into proteins*. This entire process is called *gene expression*.

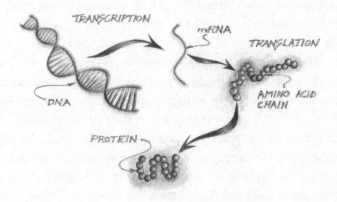

Gene transcription and translation

READY RECIPES AND BAKING RATES

I like baking cookies, and I have recipes I turn to again and again, so I keep certain recipes at hand. Cells do the same thing: they specialize in making certain kinds of proteins, depending on the type of tissue in which they reside, and so rely only on certain sets of instructions. The instructions for the majority of the other proteins that could be made are crumpled up and ignored. (Literally: most of the DNA in every cell is squished up into something called chromatin, which is the DNA wound around proteins so that it is not open for translation.) Similarly, I have a recipe for chocolate-covered bacon, but that recipe is buried in a closed book on a shelf in our pantry, gathering dust (but wait, how could that not be good?).

What genetic recipes a cell keeps at hand depends on what the cell is supposed to do. For example, among other functions, cells in the pancreas must sense blood glucose levels, those in bone must provide structural support, and those in the brain must transmit electrical signals. To perform these different functions, cells need to manufacture specific proteins. This kind of cell specialization is particularly important in growing embryos, when every new cell must eventually differentiate and decide its fate. What will it become?

Which genes—which stretches of DNA—should be kept ready and available, and which should be packed tightly away?

By adulthood most of our cells have already differentiated (a tiny minority of stem cells remains). Each of our cells retains all our DNA—the whole genome—yet produces proteins from only a small selection of genes. Inside the cells in women's facial skin, the recipe for making dark, thick hair is crumpled up in the back of the shelf—so most women have only a little bit of it on their face. But inside the cells in men's skin, the same recipe is always out, prominently displayed, to be transcribed and translated over and over again. (High T is often the culprit behind more and thicker, dark hair on women's faces, and this is something I'll explain more fully in chapter 9, "T in Transition.")

It's not simply a matter of particular genes being "on" or "off." Genes can be transcribed and translated into proteins at different rates. If protein production rises, the gene expression is said to be "upregulated," and if it's reduced, the gene expression is "downregulated."

THE BIPOTENTIAL GONAD

When embryonic Jenny was just beginning to develop inside her mother's womb, her stem cells received the usual instructions for what they should become—liver, nerve, bone, skin, etc. They received and followed directions about which genes to upregulate and which to ignore—so the usual proteins were produced that cause the cells to differentiate into the tissues necessary to build a human. The decisions about what kind of cell to become in these "unisex" tissue types don't differ much in males and females; we all need a liver and bones. But we don't all need testes, and we don't all need ovaries. So how does the developing fetus know how to make this crucial decision about whether to go down the path leading to ovaries or the one leading to testes?

During early fetal development in both sexes, groups of undifferentiated cells cluster together on the ridges of structures that will later become our kidneys. These "primordial" or "bipotential" gonads are identical in male and female fetuses up until week six, when their cells will begin to differentiate and gather together to create one or the other type of gonad. The path those cells take depends on whether genes in the DNA inside of the cells hear a loud request, in the form of high levels of a protein called SRY, which stands for "sex-determining region of the Y chromosome." The SRY protein is coded for by a gene of the same name, located, unsurprisingly, on the Y chromosome.

The sperm that fertilized the egg is the decider. Typically, each sperm carries either an X or Y chromosome, and all eggs carry one X. Whether all the embryo's cells inherit XY or XX sex chromosomes depends on whether the fertilizing sperm contains a Y or an X. The sperm decides the fate of the primordial gonads because the SRY gene is located on the Y chromosome.

At around six weeks, this Y-chromosome gene—the SRY gene—is transcribed into the SRY protein. That protein then increases (or sometimes decreases) the transcription rate of other genes on other chromosomes. Of particular importance is the gene SOX9 on chromosome 17, which is one of the first genes on which SRY acts to upregulate its activity, increasing the production of the SOX9 protein. That protein, in turn, goes on to alter the expression of other genes in the cells that comprise the primordial gonad. In this way, the SRY gene leads to the production of specific proteins inside the cells of the primordial gonad. These proteins cause the cells to take on the special

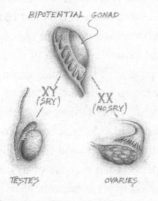

Bipotential gonad differentiation

characteristics of testicular cells (and also to repress genes that lead to ovarian cells). Ultimately, the Y chromosome and the expression of SRY, along with many other "downstream" genes, cause the cluster of cells that comprises the primordial gonad to form testes rather than ovaries.

Full ovarian development requires two X chromosomes and the expression of many different genes. But there is no analogous "master switch" like that contained on the Y for testes development. Ovaries will develop if the SRY protein is not expressed at high levels in the cells of the primordial gonad during the relevant critical period—the sixth week of fetal development. No Y chromosome, no surge of SRY, no testes formation. Ovaries can also be produced in an XY individual if SOX9 or other important genes in the "make a testis" pathway don't work as usual. In these cases, the fetus will typically develop as a female, albeit with ovaries that may not be fully functional in adulthood. An individual's sex isn't always consistent with its sex chromosomes. What matters most is the particular pattern of gene expression that leads to the development of testes or ovaries.

I'll admit it: I had no idea how any of this worked until my early thirties, when I started graduate school at Harvard and took one of my first classes, Behavioral Endocrinology, along with a bunch of undergrads (taking some undergraduate classes is standard protocol for most grad students). I had been fascinated by the differences I'd observed between male and female animals, but I'd never given much thought to how we all develop some of the most basic features that differentiate us. I think I presumed that from conception onward, only one sex destination was possible for each of us. It just seemed right that some of the clumps of cells that would go on to make up a little boy or girl were predestined to become testes or ovaries, a penis or vagina, and all the bits to which they were attached. When I learned that there are two clusters of cells that could, even at six weeks, go on to form *either* ovaries or testes, and then others that could form *either* the vagina or penis, I felt a sense of awe at the efficiency of natural selection, in making what are relatively minor adjustments to a single body plan to end up with males and females. I also felt a profound sense of connectedness to the other sex. We are made from *almost* the exact same stuff.

THE TWO SEXES

I'm throwing the terms "male" and "female" around here, but I haven't yet explained what they mean. You might think you know the answer to that, too, as I thought I did before grad school. I believed that XX and XY chromosomes defined femaleness and maleness. But that's not how it works, even though those chromosomes are distinctive of male and female mammals. XX and XY chromosomes are traits that are *features* of sex (in mammals), not ones that *define* sex.

In humans, sex is usually determined at conception, based on whether the sperm contains a Y or X sex chromosome. But sex chromosomes aren't always XX and XY. In birds, for example, males have a pair of identical sex chromosomes ("ZZ"), and females have two different ones ("ZW"). What's more, many species don't even rely on chromosomes to make males and females. In turtles and crocodiles, it's the temperature of the eggs that determines the sex of the hatchlings. And animals don't always stay one sex or the other. Every coral reef–dwelling clownfish is born male and some later switch to female. Last but not least, there are animals that are both sexes at the same time, like some snails. What do all males (or females) have in common, if not sex chromosomes? Basically it's the relative size of the sex cells or gametes. Males produce small, mobile gametes (sperm), and females produce larger, immobile gametes (eggs). Don't take that *too* literally—my son doesn't yet make sperm, but he's still male. And although my ovaries are no longer regularly producing eggs, I'm no less female than when they were cranking them out on a monthly schedule. Rather, it's the design plan for the gametes that counts.

T IS THE KEY TO BOYS

Jenny, my student with complete androgen insensitivity syndrome, had a Y chromosome and the SRY gene, so developed testes instead of ovaries. Up until about nine weeks of fetal development, when she was only the size of a grape, Jenny looked like any other male fetus. In fact, let's bring in a typical male fetus for comparison—I'll call him James. The testes of grape-sized Jenny and James did one of the things that testes do best, which is to produce lots of testosterone.

Not only do we all start out with bipotential gonads that can develop into either ovaries or testes, our early internal reproductive anatomy is also gender-convertible. Early on in fetal life we all develop two sets of primordial duct systems, but after about eight weeks, one of them degenerates while the other continues to develop. We all begin with Wolffian ducts, which can become male internal plumbing, including the epididymis (a tube attached to the testicles in which sperm mature), the vas deferens (a tube that allows the transport of sperm), and the sem-inal vesicles (which produce seminal fluid released during ejaculation), and we also have Müllerian ducts, which can become female plumbing, including fallopian tubes, uterus, and cervix.

The Müllerian ducts (female system) will degenerate if they receive

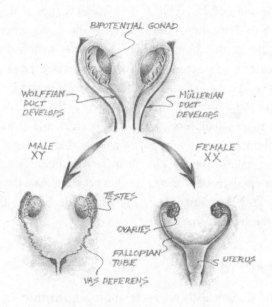

Differentiation of the internal reproductive anatomy

a hormonal signal from the testes, in the form of Müllerian inhibiting hormone. The Wolffian ducts (male system) will degenerate unless they receive a different signal from the testes—testosterone. The female duct system is the default: it will develop without any specific hormonal stimulation, unlike the male duct system.

The internal stuff is interesting, but more mind-blowing to me was learning that our external genitalia (everything you see on the outside) arise from the *very same* embryonic structures. The penis is basically a huge clitoris, and the scrotum and the line that runs down the underside of the penis are basically fused labia.

Those early structures start out looking much more like female genitalia. They have to do lots of changing to get them into male shape. When I saw all this diagrammed out, it all made so much sense (and I'm including a diagram below for you—"Differentiation of the external genitalia"—so hopefully you can experience the same sense of satisfaction as I did when I first learned about this).

Hormones like testosterone fit into their receptors rather like keys into locks, and "open the door" to all sorts of changes. If the lock is broken then the key is useless. And that's like what happened in Jenny.

It was at this point, around nine weeks, that Jenny's development started to diverge from that of James, our typical male. In James, the genital tubercle and genital folds grew into the penis and scrotum. His Wolffian ducts (precursors to male internal genitalia) developed and his Müllerian ducts (precursors to female internal genitalia) degenerated. He was humming along down the typical male pathway. In Jenny, the genital tubercle and folds grew in size but stayed pretty much as is, eventually becoming the clitoris and labia. (Although typical development of female fetuses proceeds without the need for any sex hormones, a high level of T action in female fetuses, as I will explain in the next chapter, can interfere with typical female development.) As in James, Jenny's Müllerian ducts degenerated, because they received the Müllerian inhibiting hormone signal from her testes, so she didn't form fallopian tubes or a uterus. But her Wolffian ducts also degenerated, so she didn't form a vas deferens or associated internal structures. Her vagina, which would normally connect to a uterus, just ended (a so-called blind vagina).

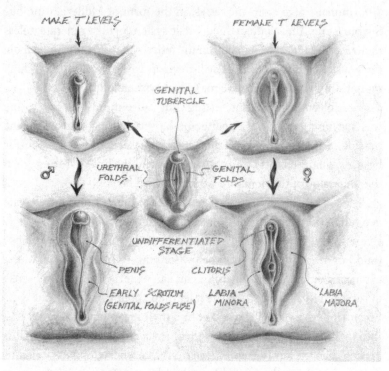

Differentiation of external genitalia

What accounted for the massively different developmental trajectories on which James and Jenny now found themselves? One teeny, tiny difference in one of the many genes on the X chromosome, a typo in one letter out of Jenny's three billion DNA bases.

KEYS NEED LOCKS

Some typos in a recipe don't make that much of a difference. Suppose you're baking chocolate chip cookies for the first time. The recipe is supposed to call for two eggs but it reads "three eggs." The cookies might still taste OK, but they wouldn't win any baking awards. That would be comparable to a mild mutation in a gene—the protein would still work somewhat, but not to its full capacity.

Testosterone's receptor—the protein lock that the T-key opens—is called the androgen receptor. As the name suggests, the androgen-receptor-lock can be opened by any androgen-key. T is the main one. The gene for Jenny's androgen receptor had a tiny typo, but the results were more consequential than misprinting "three eggs." It's as if "two cups flour" were misspelled "two cups fluor." Fluor is a mineral containing the element fluorine, and is worse than useless for baking. If you're blindly following directions, your cookies would not come out quite as expected.

Other mutations in the androgen receptor gene leave it working to some extent, more like one egg instead of two. These sorts of mutations produce partial androgen insensitivity syndrome (PAIS), where the capacity of the receptor to bind androgens varies from terrible to almost perfect. If the mutation results in only minor changes in the receptor, then the person experiences most of the masculinizing effects of androgens, developing pretty much as a typical male. At the other end of the spectrum, the mutation can completely disable the androgen receptor, and the person ends up more like Jenny. In her case, the mutation was "complete." With an androgen receptor that doesn't work at all, the body has no way of hearing the usual high T signal—"Hey! Develop male parts!"

It is high T levels in males that take the lead in masculinizing their bodies, in utero, in adolescence, and for the rest of their adult life. Testosterone is endowed with this superpower because it can upregulate (or downregulate) gene transcription. Its changing levels provide instructions to genetic recipes about whether and how often the recipes should be followed, regulating the production of various proteins in different tissues at different life stages.

High levels of T tell the "male relevant" genes to upregulate. And in adult females, the same system is at work—the sex hormones that upregulate relevant genes are mainly estrogen and progesterone. Only some genes are responsive to sex steroids, in particular those that have to do with reproductive functioning, and the secondary sex characteristics like voice, body hair, breasts, and muscle.

Steroids are a kind of biologically active chemical with four rings of carbon atoms. Testosterone (along with the other sex hormones) is one of them—it's a "steroid hormone," as I mentioned in chapter 2. Fat and water don't mix—if you put a few drops of olive oil into a glass of water, the oily drops will congregate and form their own layer. But if you add some alcohol to a glass of water, it will mix right in. Hormones are usually like either olive oil or alcohol. Steroid hormones are like olive oil. They are *lipophilic*—"fat-loving"—which allows them to float right through fatty cell membranes and into the body of the cell. Once inside the cell, they interact with receptors located there. On the other hand, protein hormones like insulin are like alcohol: *hydrophilic*—"water-loving"—and can't get inside the cell, so they interact with receptors that stick out from the cell surface.

Once inside the cell, T finds the androgen receptor and binds to it. This creates what's called the "hormone-receptor complex." Now a chain of events is set in motion. First, the hormone-receptor complex enters the cell's nucleus, where it comes into contact with the cell's DNA; specifically, with areas known as "promotor regions." Some of these promotor regions are responsive to androgens, and when activated, they upregulate the transcription of the genes they control. (See the image below to get an idea of how steroid hormones work.)

Making a female, in many ways, is easier than making a male—the external structures develop in the female direction in the absence of any hormonal signal. (Jenny had testes because their development depends on the SRY gene on the Y chromosome, not on testosterone.) Although testosterone—and working androgen receptors—is required to make a baby with a penis, estrogen is not required to make a baby with a vagina. The genes needed for that to happen will simply be transcribed in the absence of testosterone action. So Jenny's external genitalia developed as in a normal female.

That little typo in one gene turned a person who would have been like James into Jenny. Jenny's lack of ovaries and a uterus explains why she didn't get her period. But there's one last piece of the gender puzzle. Why wasn't she stuck in perpetual girlhood? How could someone without ovaries go through a feminizing puberty? Don't you need lots of estrogen to develop as a woman?

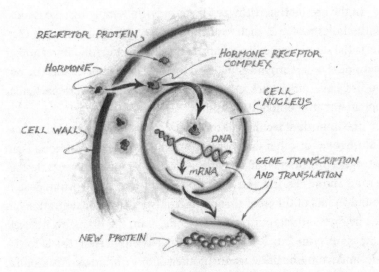

Steroid hormone action

JENNY AND PUBERTY

Yes, you do need estrogen to go through a feminizing puberty, and you also need to have a body with very low or no androgen action. This is because in a female, the masculinizing effects of even moderately high levels of testosterone will swamp the usual feminizing effects of estrogen. Surprisingly, Jenny got all of the estrogen necessary for full feminization—which is not much—from her high T levels.

Cholesterol (a type of fat) is the grandparent of all steroid hormones. In the ovaries, testes, and other steroid-producing glands and cells, there are enzymes—proteins that accelerate or "catalyze" chemical reactions—that convert cholesterol into the needed steroid hormone. Think of a branching stream fed by a lake of pure mountain water. There are a few main branches, each of which branches further into smaller streams. As the water flows from branch to branch, it changes from pure water to water with slightly different characteristics—more or less salty, more or less cloudy, depending on the kind of soil (or enzymes) the water is moving through.

In the production pathway for steroids, the supplying lake is filled with cholesterol and each branch is a place where an enzyme converts the steroid immediately upstream (the "precursor") into another steroid. (See a simplified pathway below.) Different tissues in our bodies have different kinds of enzymes, which convert particular precursors into different steroids.

The important fact here is that *in everyone, all estrogen comes from testosterone* (or other androgens). In other words, testosterone is an estrogen precursor. An enzyme called aromatase converts testosterone and other, less active androgens into estrogen. Aromatase is found in lots of different tissues, in relatively high concentrations in the ovaries and fat, but it's also in bone, skin, the brain, even testes. Like all enzymes, its levels vary from person to person, tissue to tissue, and from one life stage to another. If more aromatase is around, then more of the available testosterone will be converted into estrogen. (Bodybuilders are very aware of this and take specially altered

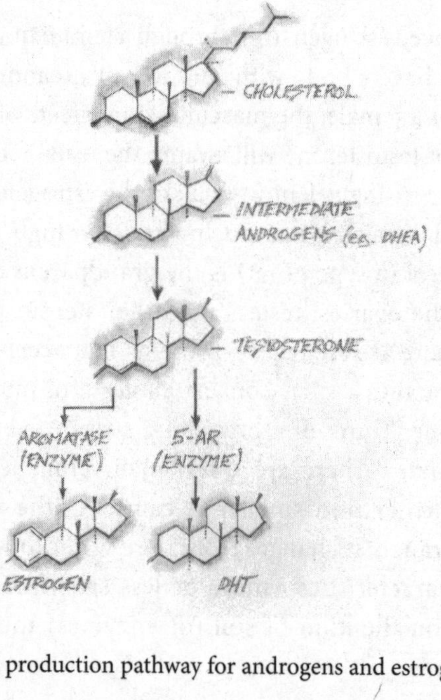

CHOLESTEROL

INTERMEDIATE ANDROGENS (eg. DHEA)

TESTOSTERONE

AROMATASE (ENZYME)   5-AR (ENZYME)

ESTROGEN   DHT

Simplified production pathway for androgens and estrogen

androgens that can't be aromatized; if they simply took high levels of T, much of it would be converted into estrogen, which would not give them the look they are going for.)

Jenny had her own estrogen production factory, in her testes. For many people with CAIS, the standard medical advice is to have their testes removed, since their presence is associated with an elevated risk of cancer. But CAIS women who have their testes removed have to take estrogen supplements to maintain their feminine physical features, in addition to bone strength. Jenny weighed the risks and decided that it made sense to have her body continue to produce her estrogen naturally, and she chose to keep her internal testicles. They have provided her just what she needed for unexceptional puberty—starting at the same age as her peers, with all the same external developments. Actually, she lit up with delight when talking about some of the ways that she was different from other girls: she had no body hair, body odor, or acne! Jenny's T was impotent as an agent of masculinization, but it worked spectacularly well to feminize her.

Jenny embraces her differences. She is grateful that she has such a loving family and medical team to guide and support her through all of her emotional challenges and medical decisions. She knows that she would not be the person she is today without her variation and is happy to have the healthy body that nature gave her.

Jenny helped me understand more about CAIS, for which I am very thankful. Jenny is thriving in her career, and planning to start a family with her husband.

SLUGS AND SNAILS

What are little boys made of?
What are little boys made of?
Slugs and snails
And puppy-dogs' tails
That's what little boys are made of
What are little girls made of?
What are little girls made of?

Sugar and spice
And all things nice
That's what little girls are made of

Jenny had an unexceptional girlhood. Given how she turned out as an adult, I can easily imagine her as a girl playing more with single friends than in groups, and enjoying playing house, not wrestling or throwing things. Ordinary girls do have some prenatal T exposure, but it's tiny compared to boys. Could this lack of T action explain why girls are made of sugar and spice, not slugs and snails?

My eleven-year-old son Griffin loves to dig up worms in the back-yard and feed them to the hungry fish in our small pond. One of his favorite ways to pass the time with his close friends is to wrestle them (the boys, not the worms) to the ground. But he also shuns team sports, has never pretended to shoot bullets out of a stick he found in the yard, and loves babies, dressing up in costumes, and making houses out of cardboard. I may be biased, but to me, he's a bit less "boyish" than many of his male peers. So his obsession with his creation, the evil genius Dr. Flergenbug, who is intent on destroying the universe, and his do-gooder nemesis, "Supersausage," has always intrigued me. Around the age of seven, Griffin began to develop an intense and complex fantasy world in which he played out this theme (with many supporting char-acters). He has spent countless hours illustrating scenes crammed with violent battles, space weapons, monsters with multiple heads and eyes dangling from antennae, aliens, and explosions. When he would proudly show me his creations, in addition to feeling pleased that he had immersed himself in a creative endeavor, I also felt per-plexed by the juxtaposition of the sweetness that I thought was my son and the evident joy he took in playing out the mortal combat and violent destruction of planets. It just did not add up. But perhaps I'm too close to my subject, because of course it makes sense. He is a boy. He tells me that his head is filled with these types of fantasies.

Am I imbuing him with stereotypical boy traits that aren't actually there, to confirm my sexist hypotheses about sex differences? Maybe my knowledge that Griffin is a boy is biasing the way I interpret his art. And that's not an idle worry. For example, in her 1985 book *Myths of Gender*, the feminist biologist Anne Fausto-Sterling describes a study called "Baby X Revisited" that illustrates how people tend to interpret children's behavior through a gendered lens:

> Experimenters told one group of subjects that the three-month-old baby with which they were to play was a boy, another that it was a girl, and then asked both groups to observe the child's behavior. All the observers ascribed sex-stereotyped behavior to the infants. One, for example, describing what he believed to be a girl (although it was really a boy) said, "She is friendly, and female infants smile more," while another found a supposed female (actually a male) to be "more satisfied and accepting" than a male child would have been.

We shouldn't rely on "common sense" or even our own observations of children's behavior to determine the extent to which boys and girls are different. Fortunately, there are numerous scientific studies investigating just this issue.

I'll describe one classic study in the next chapter. But the bottom line is that Griffin's fantasy play is representative of what boys tend to engage in: heroes battling bad guys to save the universe at great risk to themselves; the destruction of objects, homes, planets, and solar systems; and any kind of dangerous battle in which the boy can emerge as the victor (today many of these kinds of themes are played out not in fantasy but in video games). What kind of fantasy play do girls tend to act out? Those that involve relationships, romance, and domestic concerns, like getting married, parenting, going shopping, or taking care of household responsibilities. Much of girls' play, in contrast to that of boys, omits the blowing up of planets but instead focuses on coming together and finding safety after being under threat.

Kids' play is inspired by toys, and toys are brought in as part of fantasy play. The sex differences in toy preferences can be quite large and are as the stereotypes suggest. Boys are eager to play with toys having to do with transportation—like trucks and airplanes—and those that can be incorporated into battle—especially guns. The scientific literature on sex differences in toy choices is full of examples of boys, forbidden to play with guns, who become quite resourceful. I found one amusing description of a preschooler boy who picked up a Barbie doll and began to "shoot bullets" out of its head. Boys, it seems, are resistant to efforts to condition them away from battle and weaponry.

Girls, rather than battling, like to party. Specifically, they enjoy tea parties and tea sets, play furniture, stuffed animals, and dolls. Of course, girls are equally creative when their preferred toy isn't available: someone like Jenny might tuck her baby doll into the back of her brother's dump truck—the makeshift bed—and kiss her good night.

The biggest difference in play between boys and girls is the amount of physical contact with other kids. Boys do far more pushing, shoving, and hitting (usually with smiles and laughter) and delight in tackling friends and rolling around on the ground while they vie for victory in contests over who can pin the other down. Boys tend to prefer this "rough-and-tumble" play—which requires cooperation and competition—across a diverse range of cultures, from industrial populations in the United States, Europe, and Asia, to hunter-gatherers such as the Yanomami Indians of South America, the bushmen of southern Africa, and the Himba of Namibia and Angola.

NATURE OR NURTURE?

The idea that boys and girls are different in childhood play styles and interests is relatively uncontroversial—most people can see this with their own two eyes. (Of course, these are average differences, and plenty of girls enjoy rough-and-tumble play, for example, and plenty of boys dislike it. There will be more on gender-atypical play, along with its relationship to sexual orientation, in chapter 8.) What is less clear, and more controversial (a vast understatement), is the cause of the differences.

Differences in the level of prenatal testosterone could largely explain the differences in the behavior of most boys and girls, but there is an obvious alternative. The most important question when a baby is born is: "It is it a boy or girl?" The answer affects how the baby is treated from day one. We are born into a gendered social environment, with different expectations for the sexes. So perhaps differences in children's behavior are largely sculpted by social forces, and the kids help perpetuate the cycle when they grow up and have children of their own. Fausto-Sterling uses the Baby X experiment to illustrate how people's perceptions of a child's gender influence the way they treat that child:

> If the observers of Baby X believed it to be a boy (whether or not it actually was) they handed it a toy football far more frequently than they did a doll. In fact male observers never handed a football to a child they believed to be a girl.

On this alternative hypothesis, boys are not born with brains that dispose them to prefer football to sewing, or trucks to dolls. They develop these preferences because they are encouraged by their parents and other caregivers.

Let's return to Jenny and her hypothetical twin James for a moment. James liked blowing things up, Jenny liked dressing things up. Why were their play preferences worlds apart, as with most boys and girls? One explanation has little or nothing to do with differences in their brains at birth but everything to do with their bodies, which affected how they were socialized. Jenny and James differ because we all, including other kids, treat people differently according to the sex we perceive them to be. We expect boys to be tough, stoic, good with blocks and math, and we expect girls to be kind, nurturing, sensitive, and appearance-conscious. We may not mean to, but we have different habits of interaction for each sex. We push girls and boys into different kinds of activities and praise them when they behave in ways that accommodate our gendered expectations.

Since T is responsible for James's outward physical appearance (his genitalia), on this view it *is* a crucial part of the explanation of why Jenny likes playing with dolls and James likes wrestling with other boys. But the important point, according to this social hypothesis, is that T doesn't exert its behavioral effects by acting directly on the brain; instead, it works on behavior via the body. If Jenny and James were born into a gender-reversed society, where girls were expected to play with trucks and wrestle, and boys were expected to play with dolls and help to clean the house, then James would be a master sweeper while Jenny would have an extensive truck collection.

Feminist scholars and scientists are not fans of the idea that T masculinizes the brain as well as the body. In her 2010 award-winning book *Brain Storm*, Rebecca Jordan-Young argues that it is

> little more than an elaboration of long-standing folk tales about antag-
> onistic male and female essences and how they connect to antagonis-
> tic male and female natures. As a folktale, it's a pat answer, a curiosity
> killer. And the data don't fit into tidy male-female brain patterns, any-
> way. [. . .] Why keep trying to fit the data into a story about sex?

I could multiply these sorts of quotations numerous times. Gina Rippon describes the central message of her 2018 book *The Gendered Brain* succinctly: "A gendered world will produce a gendered brain." To think otherwise, says a glowing review in the leading science journal *Nature*, is "neurosexism."

So which is it? Does testosterone affect boys' brains to bias them to behave in typically masculine ways? Or are the T skeptics right, and our brains are closer to gender-neutral blank slates, to be written on by the pink and blue chalk of a gendered social world?

You might wonder how we could ever find out, since both the hormonal and social explanations can seem equally compelling. Luckily, there is a large body of research that can give us important clues.

T ON THE BRAIN

TAMAN

Taman grew up in Jakarta, Indonesia, with her parents and three siblings. Despite living in a Muslim society with strict expectations for the behavior of girls and boys, as young girl she was a "tomboy." Clad in the traditional hijab and long dress, she preferred to play outside, flying kites and climbing trees.

Around twelve, Taman began to develop what seemed to be a small penis. Other girls her age were developing breasts, but Taman's chest stayed flat. The pronoun "he" now seemed more appropriate than "she." At fourteen, the pitch of Taman's voice began to drop, his Adam's apple protruded, and his upper body widened and became more muscular. At fifteen, Taman had developed a sexual attraction to girls, and testicles descended from his abdomen into what he and his parents had thought were his labia. When Taman was finally examined by endocrinologists at eighteen, results showed that he was healthy, with XY sex chromosomes and normal T levels—for a young man. Given all this, puberty was clicking along pretty much as it should.

Taman's gendered socialization as a child obviously didn't stick. As his body changed, Taman went from thinking of himself as a girl to thinking of himself as a man. This unusual experience of puberty tells us a lot about T.

In his mother's womb, Taman's development started like a typical boy's. Under the influence of the protein from the SRY gene (sex-determining region of the Y chromosome), his bipotential gonads differentiated into testes. They cranked out boy levels of testosterone at the critical periods. But, like Jenny, who also had perfectly good testes, high T levels, and a Y chromosome, Taman failed to develop normal male external genitalia in the womb. In Jenny's case, the cause of the difference resided in the gene for her androgen receptor: it contained a mutation that gummed it up, preventing T from fitting into it, so T could not act at all. But Taman's androgen receptors worked perfectly. Still, he developed external genitalia that appeared female rather than male.

In the last chapter I told you that male genitalia develop under the influence of testosterone and that Jenny was born with female genitalia because her body couldn't respond to it. That was somewhat of a simplification. The fetus will not develop a penis and scrotum unless the androgen receptors in the precursor tissues—the undifferentiated, early reproductive tissues—receive extra stimulation, more than testosterone alone can provide. This extra stimulation comes from the more potent androgen dihydrotestosterone (DHT), which is produced from testosterone with the help of the enzyme 5-alpha reductase. Remember that estrogen, too, is produced from testosterone, with the help of the enzyme aromatase.

Like testosterone, the androgen DHT "key" fits into the androgen receptor lock but turns it more easily and stays in the lock for longer. This extra action leads to more transcription and translation of certain genes into proteins. Without DHT in utero, the external genitalia will develop for the most part as they do in females, but the internal genitalia (except for the prostate, which also needs DHT to develop) will develop as in typical males.

You can now guess what happened in Taman. His body couldn't convert T to DHT because the crucial enzyme didn't work—he had 5-alpha-reductase deficiency (5-ARD, pathway depicted below). Taman

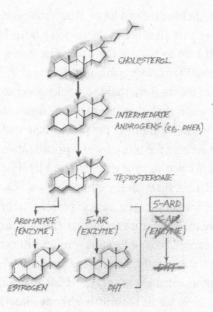

5-alpha reductase deficiency pathway

had a mutation in the gene that codes for the 5-alpha-reductase protein, just as Jenny had a mutation in the gene that codes for the androgen receptor protein. Without 5-alpha reductase, even though Taman had high testosterone, his body couldn't make enough DHT to masculinize his genitalia during his fetal development. But high levels of DHT aren't absolutely necessary to masculinize the genitals *in puberty*. High levels of T can do a decent job on their own, which is why Taman's penis later grew and his testicles descended as Taman hit his teenage years.

At the end of the last chapter, I raised the question of whether testosterone affects the fetal brain, leading boys to behave in typically masculine ways. What makes that a challenging question to answer is that fetuses with a lot of T on the brain are typically born with normal male genitalia, and so are usually sexed as boys and likely subject to potentially masculinizing social influences. How could we know if T masculinizes behavior directly through its actions on the brain or indirectly through the body, or both?

If we could somehow shoot a lot of T into the brains of fetuses and simultaneously ensure that the children were born looking just like girls, that would help to resolve the issue. The babies would then be subject to the usual feminizing social influences. If T had no direct effects on the brain, then the babies would grow up behaving like ordinary females. On the other hand, if the babies eventually played more like boys—like "tomboys" perhaps—then that would be evidence that high levels of T in utero had primed them for later masculine behavior. As you've probably gathered by now, we needn't try to get this experiment through the ethical review board: Taman's case shows that there are people who have experienced this naturally.

Of course, we can't make sweeping generalizations on the basis of studying one person. Perhaps Taman's tomboyishness was simply his unique personality as a child, or perhaps his parents' memories were biased, and they unwittingly exaggerated his masculine behavior in childhood to make sense of their little girl's bewildering transformation into a young man. But many other people have the same condition as Taman, and they have been the object of intense study—which has made clear that boys' and girls' brains are *not* gender-neutral blank slates.

THE GUEVEDOCES

In the early 1970s, Julianne Imperato-McGinley, an endocrinologist at Cornell Medical College in New York, heard of a group of "girls" in the Dominican Republic who became men at puberty. She and her research team trekked out to their remote village, accessible only by a dirt road, to meet them.

Imperato-McGinley eventually studied thirty-three such people in two villages. Nineteen of them, according to her research, had been "unambiguously raised as girls." Like the social environment in which Taman was raised, gender roles were relatively strict in the Dominican Republic. By the time children were seven or eight years old, the expectations for boys and girls diverged sharply, and they played only in single-sex groups. Imperato-McGinley reported that boys were given wider latitude to "romp and play" and were expected to help out their fathers in the fields, planting and collecting crops and tending to livestock. The girls were supposed to help their mothers with the cooking and cleaning, getting water, and bringing food to the boys and men working in the fields. While the older boys and young men attended cockfights and visited the local bars, the older girls and young women tended to stay home with female relatives, looking after younger siblings.

The villagers called the people whom Imperato-McGinley had come to see, who were raised as girls but who developed into men, *guevedoces*, sometimes translated as "penis at twelve" or "eggs [testicles] at twelve." (*Machihembras*, "first women, then men," was another name for them.) The guevedoces' stories were similar to Taman's. Living as young girls between the ages of seven and twelve, they began to realize they were unusual. Other girls began to develop breasts, but the guevedoces did not. Testicles appeared instead, and the "clitoris" began to grow into a small penis. Seventeen of the guevedoces had transitioned from living as a girl to living as a boy at or after puberty, when they began to feel sexually attracted to girls. In a later paper, Imperato-McGinley commented on the importance of pubertal changes in the solidification of gender identity and summarized these gender transitions, which occurred despite the anticipated social consequences:

> Their dawning discovery of maleness was reinforced by the development of a male muscle distribution and male body habitus at puberty, together with the onset of morning erections and nocturnal emissions. They ultimately changed gender role from female to male, a change that was enacted independent of interference from physicians and despite fear of social harassment and embarrassment. Some individuals waited until they felt confident of their ability to protect themselves from physical harm.

Since Imperato-McGinley's pioneering work, other populations with 5-ARD have been discovered in Turkey, Mexico, Brazil, Papua New

Guinea, and elsewhere. These populations share some characteristics—they are relatively isolated, gender roles are typically divergent, and families are somewhat inbred, increasing the chances of transmitting rare genetic disorders. Their isolation makes it more likely that such disorders remain undiscovered and untreated. In some remote areas, the number of people with 5-ARD is surprisingly large, but gender transitions from female to male following puberty are by no means automatic, and one large study on diverse populations found that about 60 percent of people with the condition transition after puberty. But these proportions vary according to the local culture and can be as low as 17 percent. (Of course people who did not transition to living as men may still have felt relatively masculine.)

Imperato-McGinley did not explicitly record the childhood behavior of the guevedoces, although she did mention that among the subjects of her study, "a masculine identity was evolving well before there was complete masculine pubertal development." Other accounts suggest that Taman's tomboyish interests are not atypical. For example, a BBC film crew visited the guevedoces in the Dominican Republic in 2015 and made similar observations. They interviewed Johnny, previously known as Felicita, who reported resisting having to wear a "little red dress" to school and neglecting the "toys for girls" that his parents bought for him. Instead, he wanted to play ball with the boys. They also filmed seven-year-old Carla, who was in the midst of a transition into Carlos. His mother described how the transition was no surprise:

> When she turned five I noticed that whenever she saw one of her male friends she wanted to fight with him. Her muscles and chest began growing. You could see she was going to be a boy. I love her however she is. Girl or boy, it makes no difference.

"RECKLESS EXTRAPOLATION AND FAULTY LOGIC"

Imperato-McGinley published her findings in 1974 in the prestigious journal *Science*. She was the first to identify the genes and the enzyme involved in 5-ARD and her article has been cited nearly fifteen hundred times. Five years later, in 1979, Imperato-McGinley wrote a follow-up, this time published in the *New England Journal of Medicine*. She considered the implications of research on 5-ARD and other intersex conditions for the development of male gender identity. She believed her data demonstrated that "androgen (i.e., testosterone) exposure of the brain in utero, during the early postnatal period and at puberty has more effect in determining male gender identity than does sex of rearing." She noted that "in addition to behavioral differences, androgen-induced sexual differences in brain morphology and function have been well-documented in animals." Notably, rodents. Humans, it appeared, were just another animal in this regard.

Not everyone agreed. The prominent feminist scientist Ruth Bleier, then a neurophysiologist at the University of Wisconsin Medical School, sent a blistering letter to the *New England Journal* in response. Bleier accused Imperato-McGinley of lacking "scientific objectivity," writing that "work on rodents" had been "recklessly extrapolated to interpretations of primate and human behavior despite evidence that the rodent model does not apply to primates or humans."

Bleier ended her response with this dark warning:

> My fear is that this study, like others that incorporate preconceptions, faulty logic and restrictive interpretations, will be seized on by some scientists, sociologists, psychologists and other interested parties, their liberal reluctance and protestations notwithstanding, as demonstrating that our fetal brains are irreversibly imprinted, after all, by the presence or absence of androgens: just as androgens determine our gender identity, so also, our destinies.

Bleier returned to the topic in her 1984 book *Science and Gender: A Critique of Biology and Its Theories on Women*, suggesting that her fear had already come to pass. According to her, irresponsible science like Imperato-McGinley's, which claimed that some differences between the sexes were the result of differential exposure to hormones, was now being used to make "women's subordinate position inevitable."

SEXY RATS

Let's take a closer look at some of this "work on rodents" that Bleier claimed had been "recklessly extrapolated," starting with the rat. Like many female mammals, a female rat is only interested in sex when she's able to get pregnant—when she's in estrus. And, not coincidentally, that's when males will find her most sexually attractive. During that time, she is the solicitor, and she knows what to do to get a male's attention. She's quite motivated to mate but plays a coy game with him. By doing this, both parties gain information about the other. If she performs her moves in just the right way, she signals that she's a healthy, mature, reproductively viable female. And if he responds with interest, he signals the male version of the same thing back to her. She does something that looks a little like teasing (it's actually called a "pseudo-retreat"): She runs up to him and then darts away. If he fails to follow her, she withdraws and then comes back again, and she'll repeat a few times until he begins to follow her, sniffing her behind to check her out even more thoroughly. He typically finds her approach and withdrawal behaviors very alluring. But if he fails to chase her, he's not really interested, and she and her children can do without his genes. She's motivated, but he still needs to put in some effort.

Normally, her vaginal opening points toward the ground. If he tries to mate with her when she isn't receptive, he'll get the message from her kicks and bites. In any event, he is physically prevented from mating with her unless she makes her vagina accessible to him. And that won't happen unless each animal carries out a stereotyped sequence of behaviors. He must stand up behind her, bend over, and grab the "flanks" on either side of her abdomen. Only then will she adopt the "lordosis" pose, in which she stands very still, lowering her front legs, with her back curved inward and her rear end sticking up. (Lordosis is a common mating pose among female mammals including rodents, rabbits, cats, and elephants.)

Sometimes, especially in younger animals who are just learning how to be sexual, a male will try to mount a female's head or other body parts, or even other males. Rats have to get it exactly right: if rat sex is not performed in this particular way, in this one position, no rat babies get made.

The sort of boring but satisfyingly predictable sex that is characteristic of rats and other rodents is one of the reasons that they are viewed as ideal study subjects. Sexual behavior differs sharply between the sexes, and researchers can readily manipulate the animals' bodies and environments. Today scientists can add or subtract genes and can even stimulate or block the activity of various areas of the brain, allowing them to observe the effects on behavior. Rodents and other animals have provided us with the opportunity to better understand our sexed selves by helping to generate testable hypotheses in relatively controlled environments, in ways we can't replicate in humans.

RAT HORMONE THERAPY

By the 1950s, advances in endocrinology made it possible for scientists to easily replace what castration had removed. Testosterone injections reliably restored sexual motivation and mating capacity in castrated male animals. For example, when presented with a female in estrus—usually sexually irresistible to an intact adult male rat—a castrated male is indifferent. But when that male's normal T level is restored, he again responds to that female as an object of desire, and he's able to get it on. Similarly, when a female rat's ovaries are removed—a procedure called an "ovariectomy"—she loses estrogen and progesterone, once-sexy males lose their luster, and she also loses her reflexive lordosis pose. In both sexes, gonadal hormones clearly drive the behaviors necessary to reproduce.

Researchers have long appreciated that rodents are somewhat bisexual. For example, female rats will occasionally mount, and males will occasionally show lordosis. So it was natural to think that all a female needed to more consistently show male sexual responses was the male sex hormone. But that didn't work. One couldn't simply get a female to mount another ravishing female rat by injecting the first rat with T. What was missing?

Experiments in the late 1930s had already shown that when testosterone was given to a pregnant rat, her female offspring were born with what looked like a penis. So it was clear that testosterone could masculinize fetal genitalia in the womb. But this is where T's effects in utero were supposed to end: with the development of male sexual anatomy.

By the 1950s, most researchers had come to agree that reproductive behaviors such as lordosis and mounting were most likely "preprogrammed" in development, but by genes and early experiences, not by hormones. The role of hormones was thought to be limited to eliciting these sexual behaviors in adulthood; the idea that hormones might set up the neural stage well beforehand was not taken seriously. All that was to change in 1959, with the work of a team at the University of Kansas Medical School, led by the legendary endocrinologist William C. Young.

The climate for research on these kinds of hot-button topics seventy years ago was perhaps no better than it is today. Here's Young's own account:

> Research on the relationships between the hormones and sexual behavior has not been pursued with the vigor justified by the biological, medical, and sociological importance of the subject. Explanation may lie in the stigma any activity associated with sexual behavior has long borne. In our experience, restraint has been requested in the use of the word sex in institutional records and in the title of research proposals. We vividly recollect that the propriety of presenting certain data at scientific meetings and seminars was questioned. *Counteracting this deterrent is the stimulation which has come from colleagues in many disciplines to whom we have appealed for help, and the satisfaction we have felt in seeing a picture emerge as the pieces of the puzzle have been studied and fitted together.*

Whenever I felt anxious about writing this book, I fortified myself by recalling that last line.

ORGANIZING AND ACTIVATING

Young's landmark 1959 paper called into question the standard view, that the adult brain had been organized by genes and experience, not by hormones.

In that paper, he reported on an experiment that tested whether testosterone affects neural development during critical periods in utero or shortly after birth, in ways that promoted adult male sexual behavior. If testosterone had such an organizing action, then administering T to a female fetus should predispose it to adopt male sexual behavior when given T again as an adult. The idea was that the second, adult dose would activate the areas of the brain that had been organized by T during early development.

Young and his team treated female guinea pig fetuses with T in utero by giving high levels of it to the pregnant moms, and they also removed the ovaries of the androgenized females after birth so that they could have complete control of their sex hormones.

A female guinea pig who had been exposed to T in utero was born with something that looked like a penis—so her genitalia were clearly masculinized. But what about her brain? When she was given T in adulthood, would that androgenized female act like a male and try to mount a sexually attractive female? Or, when given the hormones that induce estrus—estrogen and progesterone—would she still bend into lordosis in the presence of a sexually attractive male?

Young discovered that when a fetally androgenized female was given T in adulthood, she would behave like a male guinea pig: she would vigorously attempt to mount estrous females. But when that same female was given the estrus-inducing hormones estrogen and progesterone instead of T, she showed no interest in a normally sexy adult male and she failed to adopt the lordosis pose. Her fetally androgenized brain could not respond as that of a typical female to the usual hormones of estrus in adulthood. (And the fact that her ovaries had been removed had nothing to do with it. Non-androgenized female guinea pigs who had their ovaries removed *did* show lordosis when given female hormones.) Fetal exposure to high T had squashed the capacity for normal female sexual behavior.

Since behavior is underpinned by the nervous system (the brain and spinal cord), Young concluded that high testosterone in utero had altered the female guinea pigs' brains. If the brain is not masculinized prenatally, then the animal lacks the specialized neural anatomy that T can act on in adulthood to "activate" typical male behavior.

Young's organization/activation hypothesis was initially controversial. Ruth Bleier clearly had this in mind when she mentioned "work on rodents" in her 1979 criticism of Imperato-McGinley. Bleier did not dispute Young's research but instead claimed there is "evidence that the rodent model does not apply to primates or humans." Bleier gave no citation, though—and, anyway, she was wrong. In 1972 the basic results of Young's paper were replicated in a study of rhesus macaque monkeys, and the evidence for the organization/activation hypothesis has been steadily accumulating in humans and other animals since then, as has the T-skeptical opposition to it.

RAT RECREATION

Testosterone affects rat behavior even before puberty (and as we'll see, it does something similar in humans). Given that T is a reproductive hormone, these early effects might cause one to wonder why animals should experience this early sex programming. Why not just wait until the boy rat becomes interested in girl rats?

Rats, and mammals in general, spend a surprising amount of time engaged in what looks like a frivolous activity: playing. All that tumbling and running around seems to waste valuable energy that could be directed toward other, more sensible activities like finding food; or it could just be conserved, by resting. And small, inexperienced animals frolicking, oblivious to their surroundings, are the perfect targets for stalking predators. So why do they do it?

You might be thinking, "Well, obviously, because it's fun!" And that's right: "Because it's fun" is what biologists call a "proximate" explanation, which specifies the mechanisms (psychological, biochemical, or even social) that underpin a particular trait or behavior. But there's also an "ultimate" explanation that addresses the evolutionary history of the trait. The ultimate explanation for why rats play is because it is a way for young animals to learn and practice adult behaviors that they need to survive and reproduce. Such play behavior increases reproductive success, and so over evolutionary history it has become a prominent feature in the young of many different mammalian species.

For many vertebrate males, success in the mating arena depends on success in the dominance arena. As with other crucial adult skills, like finding food and avoiding predators, dominance skills don't magically materialize once the hormones of adolescence kick in. Juvenile play allows animals to develop adult skills. Among adult male rats, dominance pays off. High status is achieved via winning sometimes vicious fights, which cause the loser to behave submissively. Dominant males who can win fights mate more. Male rats are more aggressive than female ones (who can also be quite aggressive, such as when defending their kids). This sex difference is the product of an evolutionary history in which aggression paid greater reproductive benefits for males than for females.

We should expect to see sex differences in play in any species in which males and females benefit from practicing different skill sets in childhood, like nurturing or dominance competition, to maximize reproductive success. Human males are unusual mammals in that they provide for their young. Even so, men who provide nothing can still do quite well reproductively—but only if they are superior competitors. And women also form dominance hierarchies and benefit from competing with other females, although they don't use direct, physical face-to-face aggression often. So we don't expect men to lack the capacity to be nurturing or women to lack a drive for dominance. Yet despite the vast gulf between Rat and Man, it's unsurprising that sex differences in rat play mirror those of boys and girls in some important ways.

Like most male mammals, including humans, male rats play with other young rats more than females do. The males have their own version of "rough-and-tumble" play—biting, wrestling, and "boxing," the description of which (from a research paper on rat play), minus the "rat" and "paws" parts, seems very much like the way my own son plays:

> Boxing occurs when both rats stand on their hindpaws and push at each other with their forepaws. Wrestling occurs when two rats roll and tumble over each other. Pinning, or one rat holding [the other] in a supine position, is often the result of boxing and wrestling bouts. Pinning is used to determine dominance status among juveniles.

When male animals are experimentally prevented from playing this way, they grow up to be evolutionary losers. They are poor fighters who readily submit to intruders, have a lowly dominance status, and do not do well in the mating arena.

As we will see later in this book, the social environment affects T. But the reverse is also true, even in rats: T affects relationships and the social environment. For example, mom rats lick and groom their babies to regulate their temperatures and stimulate defecation. And the amount of T in the babies' circulation affects how much mom licks and grooms them! You might not be surprised to learn how it works: mom licks and grooms the ones with the highest T levels most. The sons with lower T levels, and the girls, get the least. And the mom's differential treatment of her kids, in turn, affects the sexual behavior of the adults they become. For example, the males who were licked and groomed the least end up taking longer to ejaculate when they grow up, and they need a longer recovery period before they can ejaculate again. The vital lesson here is that hormones can sometimes affect behavior indirectly, by affecting social interactions, which can change behavior in turn.

In many (probably all) mammals the nervous system, just like the developing reproductive system, is responsive to the lasting organizational effects of testosterone only during a limited window of time—pre- or directly postnatally (or both), depending on the species. If high levels of T are not present during those critical periods, then the lack of masculinized neural structures will reduce the likelihood that an animal will express adaptive, sex-typical behavior during the juvenile period and in later adulthood. (Just in case it's not totally clear: in humans or any other animal, this is not a problem from a moral point of view, only an evolutionary one!) It is very convenient for researchers that the critical period for the masculinization of the rat nervous system is the first week after birth (the human critical periods are during gestation and, according to newer evidence, also in the first few months after birth).

AREN'T HUMANS SPECIAL?

There is still room to argue that humans, immersed in a gendered social world, are an exception to the animal model and that sex differences in our childhood play styles have little, if anything, to do with early T on the brain. And it's true that research on rats and the like can only take us so far in understanding the basis of sex-typed behaviors in humans. Most obviously, our behavior is much more variable and flexible. Children can pretend a stick is a gun, a light saber, or a doll—or they can just play with a plain old stick. They can choose not to play, even in the presence of eager and willing companions. And sexually—well, we adult humans have a world of options available, unlike other mammals. We can have sex any time of the day or year, with an endless variety of partners and physical configurations. We can also choose not have to sex, even in the presence of an overwhelmingly attractive and receptive partner.

Unlike rats, our genes are expressed within the context of a complex cultural environment, interwoven with diverse norms and practices that significantly affect the behavior of interest. We live in a culture that often requires us, explicitly or implicitly, to conform to gender norms in one way or another. So of course we still need to test our hypotheses on humans to draw any firm conclusions about how we work.

On the other hand, rats and other animals do have much to teach us about ourselves, and it would be a serious mistake to airily dismiss research on rats as irrelevant. One of the world's experts in the neuro-chemical and hormonal underpinnings of sexual behavior, in humans and other animals, is James Pfaus, a professor of neuroscience and psychology at Concordia University in Montreal. Pfaus, who has spent much of his career investigating the relationships between human and animal sexuality, points out that the basic systems that mediate sexual responses in other mammals are largely retained in humans due to common evolutionary descent:

> Identification of common neurochemical and neuroanatomical sub-strates of sexual responding between animals and humans suggests that the evolution of sexual behavior has been highly conserved [i.e., preserved over evolutionary time] and indicates that animal models of human sexual response can be used successfully as pre-clinical tools.

A "preclinical tool" is an initial study, usually on animals, that eval-uates the effectiveness of a particular treatment, like drugs or surgical procedures, before it is tried out on humans. We've come a long way from selling extracts from crushed pig balls! Without research on rats and other animals, modern medicine would not exist.

Of course, we must be cautious not to extrapolate too broadly from studies on rats, or any other animal for that matter, to ourselves. Still, when we combine research on nonhuman animals with studies of people with 5-alpha-reductase deficiency, the overall picture supports the organization/activation hypothesis for humans. The evidence suggests that sex differences in play behavior in particular are to a sig-nificant extent due to differences in prenatal T exposure. But the data on childhood play in 5-ARD are spotty, mostly because the affected populations are so isolated. Ideally, what we want is another natural experiment where childhood play can be examined more carefully. There should be ample opportunity to investigate the play preferences of these children in detail, without relying on the perhaps biased reports of parents or the children themselves.

EXTRA T ON THE BRAIN

Congenital adrenal hyperplasia (CAH) is a rare genetic disorder, present in about one out of every fifteen thousand births, that affects the health of both males and females, but significantly affects the behavior of girls only. CAH fetuses are exposed to unusually high levels of testosterone, but in areas where families have access to modern medical care, the hormonal imbalance is typically corrected shortly after birth. (CAH is called a "disorder" here because it has effects on health that require medical treatment.) CAH girls, then, differ from unaffected girls in that they were exposed to high T levels (typically not as high as male levels) throughout their fetal development. The ways in which the behavior of these girls differs from that of other girls offers an opportunity to investigate the effects of early androgen exposure on the developing brain in humans.

The disorder is caused by a mutation in one of the genes that codes for an enzyme needed to produce the steroid hormone cortisol. Cortisol is important for freeing up energy when needed, and for life-preserving action in emergencies—the fight-or-flight response. (Today that lifesaving system is more often activated while doing things like getting stuck in traffic or giving a talk.) Cortisol is produced in the outer layer (cortex) of the adrenal glands, which sit on top of the kidneys. Mutations in different genes can cause CAH, but most often the faulty gene codes for an enzyme called 21-hydroxylase, which is part of the steroid production pathway (see a simple steroid production pathway for CAH below). This hormone converts steroid precursors into cortisol.

The lack of cortisol in the blood is registered by the pituitary gland at the base of the brain, which responds like a home thermostat when the house is cold. The pituitary works as it should, and sends out signals to the adrenal gland to work harder to produce cortisol. The adrenal gland responds as it should, and works hard to make the precursor steroids that provide the raw material for cortisol production. But without the crucial enzyme, cortisol doesn't get produced.

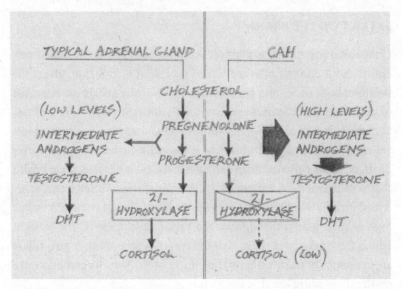

Steroid production pathway in CAH (in adrenal glands)

Instead, the hardworking adrenal gland sends all those precursor steroids down the pathway that does work, the one that converts them into androgens. More and more androgens are produced and released into the blood.

If CAH is diagnosed at birth, then it can be treated immediately with cortisol, which slows the pituitary's stimulating signals to the adrenal gland, allowing it to calm down and return to normal. The high levels of androgens in CAH have few, if any, effects on the appearance of boys, but the same is not true for girls. Female fetuses, unlike male ones, are sensitive to increases in androgen exposure, and this is reflected in the degree of masculinization of their genitalia. The amount of extra T to which girls with CAH are exposed varies, depending on the severity of the condition. Girls who were exposed to relatively low levels of T can be born with an enlarged clitoris, and if they are exposed to higher levels, the clitoris might look more like a penis. But in other ways, these girls look typically female and are almost always sexed and raised as females.

GIRLS, BOYS, AND PLAYTIME

If the organization part of the organization/activation hypothesis applies to humans, then girls who have had high levels of androgens during an early critical period should also show higher levels of male-typical behavior in childhood, similar to the effects seen in rats and monkeys.

So do CAH girls behave more like typical boys? Before getting into that question we should have average differences in the behavior of boys and girls firmly in mind. Rather than relying on anecdote or personal experience, let's start by looking at one classic experiment that elegantly illustrates differences in how boys and girls tend to interact in social groups.

Researchers divided eighty preschoolers, ages four and five, into groups of four kids each, all single sex. They told the children that they would be able to watch a cartoon through a viewer. But there was a catch—in each group, only one kid could watch the cartoon through the viewer at a time, and two others had to cooperate to run the viewer. One had to turn a crank, and another needed to keep pressing down on a light switch. The fourth just had to hang out. The researchers gave instructions to each single-sex group, and then left the small children on their own.

Apparently, the boys enjoyed the whole project more than the girls. They laughed and smiled even as they hit and shoved each other while fighting over rights to the viewer. Girls were no less competitive about the movie viewing time—they just used different, less direct tactics to get what they wanted. Girls used more "unfriendly" commands than the boys, but they were also more likely to offer their position with the viewer or the crank to other girls. Boys were more likely to use physical contact to assume a desired position. Overall, where boys used their bodies, girls used their words. Boys pushed, pulled, or hit their partners about six times more than girls did.

It's not that boys and girls don't ever use the same strategies—of course they do. Some girls are just as physical as many boys, and some boys are gentler and more apt to use verbal persuasion. But overall, as is found in study after study, boys are far more likely to compete physically to get what they want.

Before the age of about two or three years old, kids toddle around with little attention to gender. But for most kids, as soon as they come to understand that they are a boy or girl, in diverse cultures all over the world, they begin to gravitate toward their own kind. The overwhelming majority of children's playmates are members of their own sex, and this gender segregation peaks around the ages of eight to eleven. Younger kids appear to be drawn to the sex that is playing in a way that they find appealing, leading to fairly loose patterns of segregation. But as kids develop, playing with one's own sex, no matter what they are doing, becomes more important.

If a bunch of young kids are pushing each other around, laughing in a large group, moving dump trucks around the sand pit, then both boys and girls who want to do the same will be allowed to join in. If a boy wants to play house, parent the new baby, or dress up, then girls will welcome him. Preferences for larger groups and rougher, more active play, versus for smaller groups with more talking and domestic themes, drive boys and girls into same-sex groups. As time goes on and social relationships start to become more salient, it's mostly the boys who discriminate. Girls remain open to playing with boys who want to play their kinds of games, but boys tighten up their gender boundaries and rarely allow girls into their social groups. Evidently, boys have more to lose from socializing with the opposite sex than girls do.

Sex differences in children's play presage differences that have been amply confirmed in adults. We can see the seeds of sex differences in aggression, parenting, social hierarchies, and preferences for people versus things in our younger selves.

CAH GIRLS—DOES EXTRA T MAKE A DIFFERENCE?

Does that extra T on the fetal brain of CAH girls really make a difference to gendered behavior?

In studies of how the exposure to high levels of prenatal T in CAH affects children's behavior, play gets the most attention. And that's not surprising. Play is what kids like to do with their spare time, and there's no bigger difference in the behavior of boys and girls than in the way they play.

For example, in a 2005 study, researchers studied two groups of three- to ten-year-old children—with and without CAH—and gave them the opportunity to play with a variety of toys. The children could choose from toys that, according to previous research, were strongly preferred by one sex or the other, or equally by boys and girls. The "girls' toys" included a cosmetics kit, dishes, and dolls with different outfits to put on; the "boys' toys" included building logs and blocks, a gun, a tool set, and various vehicles; and the "neutral" toys included puzzles and crayons and paper for coloring. (To be clear, these toys were not selected because they were thought to be appropriate for, or naturally appealing to, one sex or the other; they were selected because they were consistently preferred by either boys or girls in previous studies.)

The boys and girls who did not have CAH—the "unaffected" children—made unsurprising toy choices. Boys spent most of their time playing with the boys' toys, and girls spent most of their time playing with the girls' toys (remaining toy-time was with "gender neutral" toys). The more interesting results were the comparisons between the toy choices made by unaffected girls and those with CAH. The CAH girls played mostly with boys' toys. They spent only 21 percent of their time with girls' toys but 44 percent with the boys' toys. In contrast, the unaffected girls showed the reverse pattern: they spent 60 percent of their time with girls' toys and only 13 percent with the boys' toys (unaffected boys spent 70 percent of their time with boys' toys and only 6 percent with girls' toys). The girls with CAH showed play preferences that were much more masculinized than their unaffected peers. (See the graphic depicting the results below.)

These results are typical of findings from the research of CAH and behavior in two main respects. First, CAH girls' play is masculinized. Second, CAH girls don't play *just like* boys, but they play *more like* boys than do unaffected girls. CAH girls' play is midway between that of typical girls and typical boys.

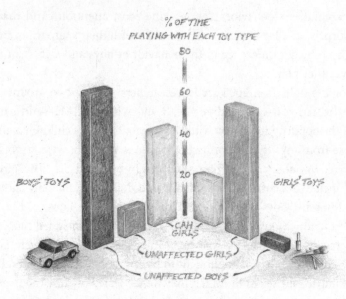

Toy preferences in CAH

Interestingly, boys who have CAH have not been found to differ from unaffected boys in their gender-typed behavior. The reason for the difference between the effects of elevated T in male and female fetuses is thought to be a "ceiling effect," where exposure to T beyond some threshold of a male-typical level has no further effect on masculinization. This is consistent with the evidence from the effects of T in adulthood, which shows that women's behavior and physiology are sensitive to small changes in T, but male behavior and physiology are not. We will see an example of this in the next chapter, when I discuss the effects of T on men's and women's athletic performance.

Over one hundred published studies since the late 1960s report on the effects of CAH on gendered behavior and have employed paradigms similar to those in the 2005 study on toy preferences discussed above. These studies confirm that the play behavior of girls with CAH is masculinized. That is, relative to girls with normal levels of androgens in utero, girls who were exposed to higher levels of androgens play more like boys. Compared to age-matched unaffected girls, CAH girls gravitate toward rough-and-tumble play; choose toys such as trucks, planes, and blocks; and have a greater preference for playing with boys.

And this tendency toward masculine behavior and preferences also extends into adulthood—CAH girls grow up to be women who are more likely to prefer male-typical professions like carpentry, that involve working primarily with things, rather than female-typical professions like teaching, that involve more interactions with people. And they make more money than their unaffected sisters!

IS IT ALL IN YOUR HEAD?

The powerful evidence from studies on CAH girls seems to seal the deal: exposure to high levels of testosterone, even before we are born, masculinizes not only our bodies but also our interests, preferences, and behaviors. As you can imagine, this conclusion generates a great deal of resistance, in the form of books and scholarly and popular articles, from those who to tend to favor explanations of sex differences that emphasize the role of socialization over genes and testosterone.

The alternative "socialization" hypothesis suggests that CAH girls are masculinized in their play not because of *direct* effects of androgens on the brain but mainly because of *indirect* effects, where their medical condition impacts how they feel about their body and how others treat them.

These girls do have lots of attention paid to their gender, to their genitals, and to their physical and psychological health, which can be uncomfortable or even traumatic. Although medical practices are changing, young girls with large, masculinized clitorises have traditionally been subject to "feminizing" surgery to reduce the size of the clitoris. And even without surgery, girls with CAH have likely experienced frequent poking and prodding of their genitals and questions about gender-related feelings and behavior. In addition, the argument goes, caregivers who are aware of the girls' medical condition may implicitly encourage masculine behavior. These social influences, in this view, explain why CAH girls behave differently from their unaffected peers.

It is true that some caregivers steer boys away from "girly" behaviors—like playing with dolls or domestic items, and toward others like playing with trucks, blocks, or science sets. But girls are generally policed more leniently, including in peer groups, where they have more flexibility with whom and how they play than boys. Boys' greater intolerance of other boys who have cross-sex interests might stem from the loss of status in the group that goes along with associating with girls.

Social influences can certainly shape kids' play styles, and it's possible that parents are an important influence. Perhaps if I had bought only "girly" toys for Griffin, like dolls and dishes, and praised him for playing with them, then he would have developed preferences for dressing up dolls and sitting them across the table for tea. Several researchers have tried to determine whether differential treatment by caregivers is somehow responsible for the masculinized play preferences in these girls. For example, kids are brought into the play lab either alone or with their parents, and toy choices are recorded under different conditions. It turns out that CAH girls play more with boys' toys rather than girls' toys, whether their parents are in the room with them or not. Furthermore, parents report that they encourage their daughters with CAH to be more feminine. The results are clear: sorry to say, parents don't matter that much.

It seems intuitively correct that social influences must powerfully separate the behavior of girls and boys, but studies like these suggest that ultimately, kids are largely oblivious to parental pressure and just do what they want. And what most kids seem to want, even at very young ages, is to do whatever their same-sex peers are doing. Somehow, that also seems intuitively correct!

Another piece of evidence goes against the idea that social forces explain why CAH girls behave differently. Measurements of T in a pregnant mother's blood are a very rough indicator of the level of T to which her female fetus is exposed. While they should be interpreted with caution, the results are generally consistent with those from the

CAH studies. They suggest that, among healthy, non-CAH girls with unexceptional female genitalia, levels of T on the high end of normal during a mom's pregnancy are associated with more masculinized play preferences in her children. And if this is right, then the outward appearance of the genitals has nothing to do with it at all.

In any event, it's unclear how having masculinized female genitals, unpleasant interactions with medical professionals, and so on, would consistently result in masculinized play styles. The T-critical scholars who favor the socialization hypothesis are right to shine a light on the complexities of the social and physical development of CAH girls and to consider how these factors may contribute to differences in behavior. But the socialization hypothesis just doesn't hold up.

EARLY T MATTERS

Evidence from experiments in mammals, from guinea pigs and rats to rhesus macaque monkeys, shows that when females are given high T in utero, their behavior is masculinized, and when males are deprived of it, their behavior is feminized. Prior to puberty, the behavior that is most affected is play style—T-exposed females play more like males. This makes perfect sense in the light of evolutionary theory. Males and females have different reproductive interests, which are served by different play styles in infancy. And research on the behavior of people who have natural differences in testosterone's actions or levels, as in 5-ARD and especially CAH, strongly suggests that humans are no exception.

The scientific principle of parsimony is useful here: all else equal, we should choose the simpler theory over the more complex one. The Greek astronomer Ptolemy tried to square observations of the planets with his theory that they revolve around the Earth by postulating weird and complicated planetary movements, or "epicycles." But the theory on which the Earth and the other planets all revolve around the sun was much simpler. We know who won that one.

Should we assume that humans are different from every other animal in which the relationship between T and masculinization of juvenile behavior has been observed, or is it more reasonable to assume that the human animal is subject to similar biological and evolution-ary forces? And if social forces alone explain why girls and boys play differently, they could presumably have produced a preference for rough-and-tumble play in girls rather than boys. On the socialization view, it is a remarkable and unexplained coincidence that social forces have exactly reproduced the kinds of differences in play that would be predicted from endocrinology and evolution—in every human cul-ture where they have been studied.

The conclusion seems inescapable: as best we can tell, T makes boy brains.

Getting an Edge

**"I AM MOKGADI CASTER SEMENYA. I AM A WOMAN AND
I AM FAST."**

From a small, remote village in South Africa, Caster Semenya surged to international fame shortly after she won the 800-meter gold medal at the 2009 World Championships in Berlin. She reached the finish line more than two seconds before the silver medal winner. Her speed, along with her muscular physique, raised questions.

At the post-race press conference, which is supposed to feature the beaming bemedaled athletes, Semenya failed to make an appearance. In her place was Pierre Weiss, secretary-general of the International Association of Athletics Federations (IAAF), the international governing authority for athletics. He confirmed the rumors that Semenya had been required to undergo "gender testing."

"But," he reassured her competitors and the curious press, "one question is clear. . . . If at the end of the investigation it is proven the athlete is not a female, we will withdraw the result of the competition today."

Weiss's announcement seemed to give license to some of the losing competitors to voice their frustrations. The Italian runner Elisa

Cusma, who came in sixth place, complained, "These kind of people should not run with us. For me she is not a woman. She is a man." The Russian runner Mariya Savinova expressed doubt that Semenya would be able to "pass" a gender test, telling her interviewer to "just look at her." (Ironically, Savinova would later be found guilty of doping and lose the 800-meter gold medal she'd been awarded at the 2012 London Olympics. The gold was then awarded to the athlete next in line, the silver medal winner, who happened to be Caster Semenya.)

Semenya's athletic triumph was soon eclipsed by sensational claims and questions in the press about her sex. *Time* magazine's website even ran an article with the headline "Could This Women's World Champ Be a Man?" Many of these stories featured pictures of Semenya just after she crossed the finish line—triumphantly flexing her arms, fists clenched by her head—that emphasized her muscularity.

In the wake of this very public controversy about her intimate medical history, Semenya appeared a month later on the cover of the glossy South African *You* magazine. She was almost unrecognizable. Her nails were long and painted purple, her hair had been styled in a bob of soft curls, and her face had been expertly made up. She wore a black dress with a long, bold gold chain draped around her neck, and so many gold bracelets that they reached halfway to her elbow. The colorful spread inside the magazine continued the theme of a feminized Semenya, happy in high heels and replete with sequins.

Clearly Semenya wanted to take some control of the narrative, but Weiss, the IAAF representative who seemed to have skipped the sensitivity training part of his job, weighed in again, offering hungry reporters a few tantalizing tidbits about the results of her gender tests: "It is clear that she is a woman, but maybe not 100 per cent. We have to see if she has an advantage from her possibly being between two sexes compared to the others."

This alleged advantage was Semenya's high level of testosterone. The IAAF soon announced that she would need to sit tight while the authorities determined whether she would have to lower her T level—and if so, by how much—to be able to compete as a woman. Eventually Semenya was allowed to return to competition with her T level unchanged, and her success on the track continued until 2018. That was when the IAAF introduced new regulations applying to athletes with differences of sex development (DSDs)—which Semenya, along with her supporters, claimed were meant to target her in particular. The regulations specified that if she did not lower her T (which would have required taking drugs), she would be banned from middle distance track events in the women's category. They defended their regulations, saying that there are some areas, including sports, where "biology has to trump gender identity." As of this writing, Semenya has refused to lower her T levels, saying that she is a woman and that she should not have to alter her natural body to compete as one.

Does higher T give athletes an advantage? Veronica Ivy (who at the time was known as Rachel McKinnon), a transgender woman and activist, former philosophy professor, and a two-time Masters world champion cyclist in the women's division, argues that the testosterone regulations are based on flawed science, and expressed that view in a 2018 interview with a cycling magazine. The idea that "the more endogenous testosterone you have . . . the better you are" is, Ivy claimed, a "myth." (*Endogenous* testosterone is T that is produced by the body, as opposed to *exogenous* testosterone, which is T introduced externally, for instance by injection. Your body can't tell the difference between the two types of T in your blood.)

SOFIA AND SAMUEL (AND SERENA)

One thing that is not in dispute is the large sex difference in sports performance. We can rely on one athlete in particular not to sugarcoat the facts: seven-time Grand Slam winner John McEnroe. In a 2017 interview on National Public Radio, host Lulu Garcia-Navarro asked McEnroe why he had called Serena Williams the best "female player" in the world. Why hadn't he called her the "best player in the world, period?" "Because," replied McEnroe, "if she played in the men's circuit she'd be like 700 in the world." He went on to call her an "incredible player" but reiterated that she'd have no chance against the best men.

Predictably, McEnroe was denounced as a sexist, and a clearly irritated Williams fired off a couple of stinging tweets in reply. Still, even Williams had expressed agreement with the underlying sentiment. When Williams appeared on *The David Letterman Show* in 2013, Letterman asked her what would happen if she played one of the top men's tennis players. Williams explained how she would lose, badly, to the reigning world champ, Andy Murray:

> Andy Murray, he's been joking about myself and him playing a match. I'm like, "Andy, seriously, are you kidding me?" For me, men's tennis and women's tennis are completely, almost, two separate sports. If I were to play Andy Murray, I would lose 6–0, 6–0 in five to six minutes, maybe 10 minutes. No, it's true. It's a completely different sport. The men are a lot faster and they serve harder, they hit harder, it's just a different game. I love to play women's tennis. I only want to play girls, because I don't want to be embarrassed.

This chapter will help to explain why Andy Murray would crush Serena Williams in a tennis match. It's about how sex differences in testosterone contribute to men's superior performance over women in most sports. The chapter will *not* weigh in on who should be allowed to compete in the female category. Instead, I want to dig into the science that helps to inform the answers to such questions. In particular, what is the evidence that testosterone really confers an athletic advantage?

Let's start with those halcyon days when the genders are nearly equal on the track—childhood.

Meet two fictional fraternal twins: Sofia and her brother Samuel. Sofia and Sam are quite ordinary. In elementary school, Sofia loved math, gossiping with other girls, playing Little League, and baking with her mom. Sam enjoyed drawing comics, playing the piano, and having laughter-infused wrestling matches with other boys. Despite being boringly average, Sofia and Sam are unusual in one respect. As they got older, they started to obsessively compete against each other in sports. At age six they ran against each other in the 30-meter dash. At ten they competed in the mile and the 25-yard freestyle swim and attempted to outthrow each other in the javelin. As they passed through the turmoil of adolescence, their passion for sports only intensified, and now in their twenties they are running marathons together, pole vaulting, and weight lifting—the works. No prizes for guessing who wins these contests of sibling rivalry.

Yet up until around age ten, neither Sofia nor Sam could consistently outrun or outthrow the other, and neither had bragging rights in any sport. It wasn't until Sam hit puberty at around age twelve that he started to really pull ahead. At fifteen, he was more than four seconds ahead of Sofia in the 30-meter dash. He could throw much farther, and with more accuracy. (The difference in throwing is vast. As one researcher put it: "Nearly every boy by age 15 throws better than the best girl.") Long-distance swimming is one of the very few sports where Sofia has the advantage.

Sex differences in testosterone levels and sports performance
(jumping events, swimming, and running) across puberty

MIND THE SEX GAP

I ran a marathon when I was thirty and it took me four hours to complete, a pretty good time for a woman of that age. I ran another at forty and it also took me four hours, an even better time considering my extra decade. I intended to run my third marathon when I turned fifty, and trained away, feeling fantastic, running at the same pace I had for my earlier marathons. I still had it going on. That is, until my fifteen-mile-long run. Various body parts started to hurt, a lot. But I pushed through it (generally a bad idea). I kept training, but eventually I had to stop running altogether because the injuries were piling up. (I've since resumed running, but more responsibly.) Nothing like this had happened before. Why now, at age fifty, did I have to be so much more careful?

Because a fifty-year-old body is not the same as a thirty- or forty-year-old one, no matter how well you treat it. In many sports, segregating people into classes like age and weight gives each participant a more realistic chance of succeeding. Sporting performance varies even more with the number of usable limbs and the capacity to see, process information, coordinate muscles, and so on. This is the reason for the Paralympics—it allows people with physical or mental traits that would otherwise disadvantage them to compete on a playing field that is somewhat leveled. But the most basic categorization, in almost every elite sport, is sex. And, until recently, the reason for that separation was obvious and uncontroversial.

Women's world records are consistently about 10 percent lower than those of men. For example, the marathon world record for women is two hours and fourteen minutes, around twelve minutes slower than the world record for men, two hours and two minutes, set by the Kenyan Eliud Kipchoge when he was thirty-three.

This sex gap in performance means that in many events, thousands of male athletes are ahead of the very best female. In 2019, about *twenty-five hundred* men, almost one-third of the total number of men competing worldwide in the IAAF 100-meter event, beat the fastest women's time. Without segregation, it's not just that men

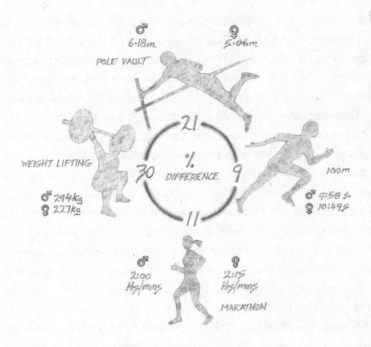

POLE VAULT
♂ 6·18m   ♀ 5·06m

WEIGHT LIFTING
♂ 294kg
♀ 227kg

21
30   %
      DIFFERENCE   9
11

100m
♂ 9:58 s
♀ 10:49 s

MARATHON
♂ 2:00 Hrs/mins
♀ 2:15 Hrs/mins

Sex difference in world records

would win—women would never even qualify for the competitions in the first place.

Veronica Ivy, the Masters world champion cyclist in the women's division, is more optimistic: "We've seen that the gap in performance between elite men and women is closing in every sport. As the men are improving and new records are being set, the women's records are being set faster. The gap is closing. It's misleading to take the current gap and say that will always be the case."

Here Ivy is mistaken: whatever the cause, the gap is not closing. After the implementation of Title IX (banning sex discrimination in education programs receiving federal funds) in the US in 1972, women's performance did increase relative to men's, but that lasted barely a decade. From 1972 to 1980, the sex differences in Olympic trial times went from 17 percent to 13 percent (running) and 13 percent to 11 percent (swimming). In the last forty years, the gap hasn't budged.

Other T skeptics appear to think that women do not perform as well because they don't try hard enough. In 2018, the BBC radio program *Woman's Hour* aired a segment on transgender participation in athletics. The issue under consideration was whether transgender women (people born male who identify as women, discussed in chapter 9) should have the right to compete in women's sports. One of the guests, a psychologist named Beth Jones, was dubious that testosterone was a factor: "There is no robust scientific evidence to say . . . testosterone impacts directly on athletic performance." She suggested that "in years to come" it might be feasible to abolish sex categories in sport. When the interviewer objected that this would mean the end of women's participation at the elite level, Jones speculated that "women cap their capability psychologically because they are competing against other women. If they feel they're then competing against men—perhaps they would up their performance and be competing on more of that level."

Why should some scholars be motivated to deny what most endocrinologists and sports scientists think is glaringly obvious? Perhaps because it's true that we have been getting T wrong, and its supposed powers are actually mythical. And there is plenty of bad science out there.

BAIT AND SWITCH

One sort of argument used by the T skeptics can seem persuasive because it's backed up by high-quality research in endocrinology. Here's an example, from a 2019 article in the *Washington Post* by Rebecca Jordan-Young and Katrina Karkazis, "Five Myths About Testosterone." *The more testosterone, the better the athlete* is myth number five:

> No study has ever concluded that you can predict the outcome of speed or strength events by knowing competitors' T levels. And while T does affect parameters related to athleticism, including muscle size and oxygen uptake, the relationships don't translate into better sports performance in a clear-cut way.

The authors then go on to describe findings from two studies, one based on an Olympic weight-lifting competition, and one on elite track-and-field athletes. Each study showed that those with the highest T levels didn't always win. For some women, in some events, higher T predicted better performance. But for women in other events, T either had no relationship to performance, or low T was actually associated with an advantage. And similar relationships were found among men. These kinds of findings, Jordan-Young and Karkazis say, "cast doubt on the thesis that T is a master key to performance differences in sports."

They are right in one way—T is not the master key to performance differences *within each sex*. Many studies show that among elite athletes, T level (when it is within the normal range) is not related to performance, *within sex*. That is, if we measure the T levels of healthy, typical male (or female) athletes, we will not find that the men with the highest T levels *relative to the other healthy men* are always the best at their particular sport. (In endurance sports, in which some of the best men have relatively low T levels, we will find something like the opposite result.) But the conclusion that T levels don't explain differences *between* sexes is the result of a subtle bait and switch. The scientifically backed claim, that T doesn't always predict sporting success *within* males, or *within* females, has been replaced by a similar-sounding but unsupported claim that T doesn't explain sporting differences *between* males and females.

As I will show, all the evidence points to the same conclusion: a male level of T, in puberty and adulthood (with possible contributions from prenatal T), is the master key for superior performance in most sports.

T LEVELS

The issue of a person's T level is actually much more complicated than it might seem. First, nobody really has one "T level" or even a "Monday T level." T levels vary over a person's lifetime, and also over the course of the day, being highest in the morning and lowest at night.

Of particular relevance to athletes, prolonged physical exertion can also reduce T levels temporarily. And "doping" with androgens can lead to long-term reductions in endogenous T, because such high T levels in the blood tell the testicles to take a break from steroid (and sperm) producing activities.

Second, the T level that studies report is usually measured in either saliva or blood. Remember that testosterone, like all steroids, doesn't mix well with water—it is *hydrophobic* (water-fearing). So in order for it to get around in the watery blood, it is carried by *hydrophilic* (water-loving) proteins.

About 98 percent of T in the blood is stuck or "bound" to a hydrophilic carrier protein, and that protein can't pass through cell membranes. Since that bound T can't get inside cells to interact with their androgen receptors, it doesn't have biological action (that we know of). The other 2 percent of blood-borne T is "unbound" or "free." Free T can pass through cell membranes, interact with the androgen receptor, and affect gene transcription. In this way, it can affect the functions of the body or brain. When testosterone levels are measured in blood, the results generally reflect either the total amount in the blood, which is the bound plus the free T, or just the free portion. It depends on the type of test.

The testosterone in saliva is mostly free (unbound, biologically active), so salivary T levels are much lower than T levels in the blood. This needn't cause any confusion if we are careful only to compare apples to apples.

If studies tell us that there isn't much of a sex difference in T, then T can hardly account for men's advantage in sports. So first we need to sort out whether there is, in fact, a large sex difference in T levels, and to do that, we need to understand the different ways that T levels can be measured and reported.

Measuring your T isn't like measuring your height. Rulers and tape measures are all easy to use and deliver pretty much the same result. But there are different methods for measuring T, and they all involve sophisticated technology. What's worse, the different methods don't even come up with the same numbers, even when comparing apples to apples!

The least expensive and most common method is called radioimmunoassay (RIA), which can measure T in saliva or blood. But you
get what you pay for. RIA has enough problems that we stopped using
it to measure female T levels years ago in our Reproductive Ecology
lab at Harvard (and many other research labs have done the same).
First, different kinds of test kits can yield vastly different results.
And second, in women, immunoassay tends to pick up other, weaker
androgens that are similar in structure but don't have the potency of
testosterone in tissues like muscle. This "cross reactivity" with other
steroids can drastically inflate women's reported T levels! This isn't
a problem in men because their T levels are so high in relation to
weaker androgens that the cross-reactivity makes no difference.

As a recent study examining the accuracy of T measurements based
on RIA concluded, "This tendency to inflate very low concentrations
of testosterone creates a substantial impediment to accurately assessing women's testosterone." Another study that evaluated the accuracy
of RIA put its limitations more bluntly: "Are assays that miss target
values by 200–500% meaningful? Guessing would be more accurate
and additionally could provide cheaper and faster testosterone results
for females."

These problems don't mean that we should toss out all the results
on testosterone and behavior in which T levels have been measured
with radioimmunoassay—for one thing, they are relatively reliable
for male T levels. But it does mean that we should be cautious about
claims made about sex differences in T levels that are based on studies
that use RIA to measure women's T.

Sari van Anders is a professor of psychology, gender studies, and
neuroscience at Queen's University in Kingston, Canada, who spe-
cializes in "social neuroendocrinology, sexuality, gender/sex & sexual diversity, and feminist and queer science." She has claimed in her
scholarly work that "in reality, testosterone levels *overlap considerably*
between women and men." (She is not the only scholar to say so.) Interviewed recently for *Discover* magazine, van Anders reiterated the claim
that there is no gender "binary" in T levels and asked, "What would be
the goal of making that binary? For science, there isn't one. Usually,

[the binary] is there for political reasons." Van Anders's work is interesting and innovative, and it challenges traditional ways of thinking about and studying sex and gender, partly by including people with underrepresented sexual identities in her studies. (I've relied on some of her research on men and parenting for this book.) But here, the evidence is not in her favor.

So what does the evidence from more reliable measurement methods show? Mass spectrometry ("mass spec") is the gold standard of T measurement and is increasingly used by clinical and behavioral researchers in endocrinology. It is the only method used by antidoping agencies, which require a high level of accuracy for both males and females.

The most recent, comprehensive, and rigorous study on adult T levels was led by an Australian endocrinologist, David Handelsman. Handelsman is one of the top experts in the world on androgens, their function, and their measurement in athletes.

Handelsman and his collaborators surveyed the scientific literature and compiled a list of studies on testosterone levels in adults, all of which relied on mass spectrometry. This kind of study, a "meta-analysis," is vital for understanding a body of scientific literature, and what science as a whole has to say about a particular question. Rather than providing one set of data from one study, which could be unreliable for a host of reasons, a meta-analysis consolidates, compares, and evaluates data from many different studies. Results that are consistent across studies provide strong evidence for support of a particular hypothesis.

The researchers evaluated thirteen studies, published from 2005 to 2017, that met their high standards for eligibility. Only studies that reported on T levels sampled from blood (rather than saliva) were included, since this method is the most accurate, particularly for women. None of the subjects in these studies were known to have any health problems that would affect their testosterone levels, so were assumed to be a representative sample of healthy men and women, aged about twenty to forty years old. The number in each study ranged from a low of twenty-five to over fifteen hundred people, and most tested the T level of over one hundred subjects, which is large for any study of this kind.

There was widespread agreement in results across the studies, which confirms the accuracy of the T measurements. In particular, there was very little difference in the high and low end of the average T ranges for each sex. This consistency provides a strong starting place to evaluate the extent of the overlap in T levels between the sexes. Based on the evidence from T measurements from the various independent research labs, Handelsman concluded: "Circulating testosterone in adults has a strikingly non-overlapping bimodal distribution with wide and complete separation between men and women."

We already saw a "bimodal" distribution in the first chapter—the distribution of adult height. A bimodal distribution has two peaks. And in the case of height, the male and female distributions look like two mountains with wide, overlapping bases. In other words, male and female heights significantly overlap, since some men are shorter than many women and some women are taller than many men. But with testosterone levels, that bimodal distribution has a *wide and complete separation*," like two mountains separated by a vast plain, as you can see in the figure below. In other words: a binary.

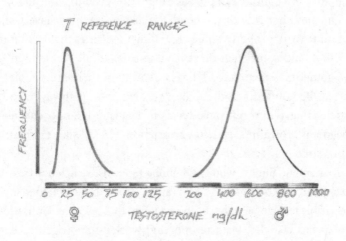

Blood T levels in healthy young men and women

The claim that the natural T levels of healthy adult males and females significantly overlap is simply not supported by the scientific evidence. Rather, there is robust evidence of considerable separation. In people without DSDs, serious disorders of their endocrine glands, or other rare conditions that have a large impact on T levels, men's T levels are ten to twenty times higher than those of women.

MEDICAL CONDITIONS AND T

The graphic above excludes people with medical conditions that result in unusually high, or low, levels of T. Polycystic ovarian syndrome (PCOS) is one example. It is a disorder of the ovaries, usually associated with ovarian cysts (fluid-filled sacs) and the overproduction of testosterone and other androgens. T levels in women with PCOS are on the high end of normal but may be above the normal female range. (That high T frequently leads to the development of masculine traits like increased facial hair and acne.) PCOS afflicts anywhere from 5 to 20 percent of women of reproductive age, making it the most common reproductive disorder in women of this age group. A much rarer condition is congenital adrenal hyperplasia (CAH), as we learned in the previous chapter. If these women remain untreated, their adrenals continue to produce high T levels into adulthood.

On the other side of the gender tracks, some men have testicles that don't work properly, some are taking medicines that inhibit their function, and some men don't have testicles at all. These men have low, sometimes near-zero, T levels. And then there are people who started life going down the male developmental pathway, who have testicles that pump out male levels of testosterone but who do not develop as typical males. Jenny's variation, CAIS, is an extreme case of this sort.

Remember Taman, who had 5-alpha reductase deficiency (5-ARD)? He appeared to be a girl and was raised as one, but developed a penis and visible testicles at puberty and began to live as a man. Unlike Jenny, Taman has testicles, testosterone, *and* working androgen receptors, but lacks the crucial enzyme, 5-alpha reductase, needed to convert testosterone into the more potent DHT (dihydrotestosterone). A substantial proportion of people with 5-ARD continue to live as women through and after puberty (and in some cases, their testicles never descend). And among elite athletes with DSDs who compete as women and have XY sex chromosomes and high testosterone, 5-ARD is common.

Handelsman's meta-analysis that confirmed the gender-binary in T levels was based on large numbers of typical and healthy adult males and females. But perhaps that binary would disappear when people with DSDs or common medical disorders were included in the analyses. To answer that question, another research team, led by Richard Clark, a member of the board of directors for the U.S. Anti-Doping Agency, conducted a similar literature review that included people with PCOS, CAH, and DSDs like 5-ARD and PAIS (partial androgen insensitivity syndrome, in which the androgen receptors have a degraded capacity to respond to T).

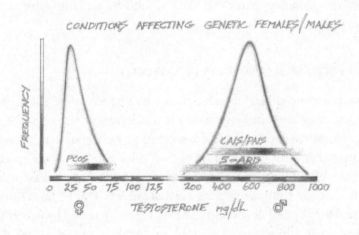

Blood T levels including people with atypical conditions

The graphic above illustrates the T ranges of people (with XX sex chromosomes) who have PCOS and people (with XY sex chromosomes) who have CAIS/PAIS and 5-ARD. Clark and his team found that people with PCOS—who have ovaries—had T levels that were on the high end for females but that still did not even reach into the bottom end of the normal male range. They also found that people with 5-ARD and PAIS (who have not had their testes removed) had testosterone levels largely within the male range.

When people with XX chromosomes have a T level within or even exceeding the male range, that high T is likely to be produced by ovarian or adrenal tumors. Apart from those serious medical conditions, the only natural way to have adult male levels of T is to have testicles and experience male puberty.

Of course, these differences in T levels alone do not show that T is responsible for differences in athletic performance. So back to our fictional twins, Sam and Sofia: perhaps Sam grew bigger, stronger, and faster than Sofia because he has "athletic genes" that are unique to males. Or perhaps there's some other biological explanation. Or perhaps his parents force-fed him Wheaties. But masses of evidence from endocrinology show that Sam's advantages are largely due to T. And it all starts in puberty.

BOYS INTO MEN (AND GIRLS INTO WOMEN)

Sam experienced heightened exposure to T at four different times. He had a big boost in T in the womb, which masculinized his reproductive system and brain. He also enjoyed a large elevation in T levels shortly after birth (the purpose of this "mini puberty," as it's called, is not well understood). His T soon dropped to baby Sofia's rock-bottom levels, until the next big burst at puberty. During this critical period for physical development, Sam's T level increased by twenty to thirty times, while Sofia's T budged up slightly. And then there's the rest of his life—Sam's T levels peaked at around age twenty, plateaued for a few years, and then slowly declined. In Western populations, the average decline is about 1.2 percent per year after age forty. (But this decline appears to be due largely to being outliers in the world of men's T. The norm in much of the rest of the world, in less industrialized and small-scale societies, is for men to have lower levels overall, and with relatively little decline throughout adult life.)

Puberty is a critical period for the development of traits that contribute to athletic success. In both of the twins, it was kicked into gear by the hypothalamus, an evolutionarily ancient, almond-sized structure deep in the vertebrate brain that acts as a kind of bridge between the nervous system and the endocrine system. The hypothalamus began to send pulsatile signals to the pea-sized pituitary directly below it, in the form of gonadotropin-releasing hormone (GnRH). In response, the pituitary began to send its own signals—called luteinizing hormone (LH) and follicle-stimulating hormone (FSH)—through the bloodstream down to Sofia's ovaries and Sam's testes. This system (depicted below) is called the hypothalamic-pituitary-gonadal (HPG)

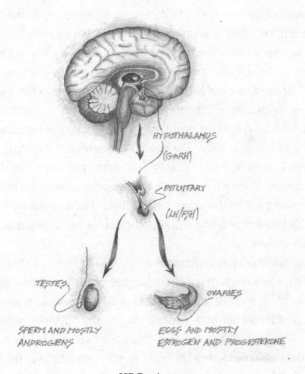

HPG axis

axis and is front and center in the control of the production of our sex hormones, along with eggs and sperm.

Sofia started puberty at a typical age for females, around age eleven. Soon after her ovaries received the signal from her pituitary gland, her pubic hair grew, breast buds appeared, pimples started to pop up, and her growth in height accelerated. She got her first period a year later. Sam started puberty at twelve and a half (boys start one to two years after girls), also developing pubic hair and oily skin. In addition, his penis and testicles enlarged, his voice began to crack and drop, and eighteen months later he experienced his first ejaculation. Sam was soon taller than his mom. Although Sofia's growth spurt started before Sam's, it also ended earlier.

Over the course of the next three to four years, these kids became teenagers—no longer children, not yet men and women, but with bodies that are capable of creating a new life. For reproduction to be successful, Sam and Sofia's bodies must develop very differently. First and foremost, they need to create little vehicles—gametes—that can get their genes into the next generation. So the gamete production factories, the testicles and ovaries, come on line—but just as a factory that produces widgets is a waste of money unless the widgets can be sent to customers, sperm are useless without a system to deliver them. Similarly, eggs are pointless without a place to house and nourish the developing baby. The rising sex hormones ensure that sperm and eggs are paired with a body that is ready to use them. Evolution has ensured that when sperm and eggs aren't available, then energy isn't wasted on all the other physical features that aid reproduction. During those times, like childhood or older age in women, sex hormones are low.

The secondary sex characteristics, bodily features that appear at puberty, like breasts or a deep voice that make it easier for us to identify people as male or female, are the product of sex hormones. Sofia's increased fat on her hips, breasts, thighs, and buttocks, as well as her wider hips, are more examples. Wide hips are needed to accommodate the passage of a small human. Sofia's extra fat stores the energy to grow another human from scratch inside her own body, then to nourish it with milk that she's produced once it's on the outside. Sam might be able to get away with just producing sperm, but his extra muscle and larger body size help him compete for the status or resources that will help him to attract mates and provide for his offspring. So in addition to developing his reproductive system, high T helps Sam to grow a bigger body than Sofia's, including stronger and longer bones and increased muscle mass.

In both Sam and Sofia, their sex hormones act like the manager of a team of hormones, including growth hormone (you can guess what that does), insulin, insulin-like growth factor 1 (IGF-1), thyroid hormone, and others. These hormones all work together to transform Sam and Sofia's bodies in ways that support reproduction. The sex hormones lead the way, but they couldn't do their jobs without the help of the others on their teams.

Shortly after the onset of puberty, Sam can reliably beat his sister in an arm wrestle. Along with growth hormone and IGF-1, testosterone promotes muscle protein synthesis. T influences stem cells—those that haven't yet decided whether to become fat or muscle—to take the muscle route, and it actively discourages them from turning into fat. Higher T in Sam also enlarges those muscle fibers, resulting in stronger, larger muscles.

Sofia's higher estrogen and lower T means that, compared to her brother, she converts more of her consumed energy into fat rather than muscle. Across puberty, while both Sam and Sofia put on fat, Sofia's rate of fat accumulation is twice as fast as Sam's! When their bodies cease their pubertal development and have stabilized in their adult forms in their late teens, Sam's lean body mass (everything except fat) is about one and a half times that of Sofia's. While that fat will fuel her ability to produce children, when it comes to sports, for the most part, the extra fat just means she has more dead weight to carry than her brother. It's as if Mother Nature has come along and strapped a few bags of flour on Sofia, giving Sam nothing to carry. This means Sofia is not going to win a pull-up contest against her brother.

It's worth taking a moment here to emphasize that Sofia is not "biologically destined" to lose that pull-up contest. Differences in all kinds of habits, even in words of encouragement from the parents, might allow Sofia to beat her brother in any number of sports. If Sam blobbed out on the sofa all day, playing video games and eating cupcakes, and Sofia spent her time training and eating healthily, she'd improve her prospects immensely. There's no question that environment, upbringing, culture, and personal habits affect what people are capable of athletically. But all else being equal, Sam is the one likely to enjoy the sporting advantage. And not just because he has more muscle; bone differences matter, too.

Bones have receptors for both estrogen and testosterone, and both are involved in growth of the long bones; but estrogen plays the dominant role, in boys as well as girls. So how is it that Sam's bones grew more in puberty than Sofia's? Remember that *all estrogen comes from testosterone*, with the help of the enzyme aromatase. The important role of estrogen in bone growth for boys went unappreciated until the 1990s, when a small number of boys with the very rare disorder of aromatase deficiency were studied. A mutation in the gene for the aromatase enzyme meant that those boys could not produce estrogen— without aromatase they were unable to convert testosterone into it. These estrogen-deficient boys, and the men they became, had similar bone issues as the eunuchs. These boys were tall, with long limbs and weak bones (along with some other metabolic issues, demonstrating the importance of estrogen in male growth and health). Doctors initially attempted to treat the condition with testosterone injections, but T had no effect (the boys had plenty of T to start with). When estrogen was given instead, their bone density normalized and their bones finally stopped what had essentially been an elongated period of childhood growth.

Elevated estrogen (from testosterone in boys) promotes bone growth for most of puberty, but it's also estrogen that stops bones from growing at the end of puberty (androgens also act in bones to promote growth, but their actions are not as important as estrogen). Girls may briefly overtake boys in height because they start puberty a year or two earlier, so they get a head start on their growth spurt.

But they also end their growth spurt at a younger age, around fourteen or fifteen. That extra year that boys spend in their childhood growth means that they enter puberty taller than the girls were when they started. That head start on height contributes to the taller adult height that boys reach at the end of their puberty, at around age sixteen or seventeen. In both sexes, the growth spurt ends when estrogen reaches a high threshold toward the end of puberty and causes the growth plates of their long bones to be sealed.

Sam's larger and stronger bones are a result of his increased muscle and higher T levels. In puberty, bone architecture is especially sensitive to mechanical load and develops in response to it. And the muscles of pubertal boys apply greater load to the developing bones than do those of girls. Stronger and bigger muscles constantly pulled on Sam's bones, causing them to respond by increasing mineral density and diameter. All of these effects—lengthening, enlarging, and strengthening—are, for the most part, permanent. Sam would have a bone strength advantage (not to mention height) over Sofia even if he reduced his T to match hers in adulthood.

Plus, the elevation in T at puberty increases Sam's hemoglobin levels, as it does in nearly all adult mammals. (Men's levels are about 12 percent higher than women's.) Hemoglobin is a protein inside red blood cells that carries oxygen from the lungs to working muscles, fueling their action and increasing endurance, among other benefits. (This effect of T is not permanent—Sam or Sofia would be able to manipulate their hemoglobin levels in adulthood by changing their T levels.)

In addition to these T-induced pubertal changes, Sam will benefit from the day-to-day effects of T throughout the rest of his life. High testosterone will keep his primary sex characteristics—penis, testicles, and internal reproductive structures—in working order. (Low testosterone in men can cause erectile dysfunction, testicle shrinkage, and low sperm production. Later, we'll explore what happens when adult male-to-female transgender people, who may not see high T as a benefit, suppress their testosterone.) But high T will also maintain Sam's secondary sex characteristics, most relevantly his bone strength and increased muscle mass. In addition, his high T will keep his hemoglobin levels pumped up, giving him greater aerobic power. And with T continuing to keep his fat low, along with his intense training, Sam's athletic dominance over Sofia is assured.

T, MUSCLE, AND T SKEPTICISM

You may think it glaringly obvious that testosterone builds muscle—just look at athletes like Barry Bonds and Mark McGwire, before and after "juicing" with anabolic steroids! What's more, when researchers castrate animals, disable their androgen receptors, or reduce testosterone, muscle changes occur in the predictable way. Experimentally manipulating testosterone levels in humans shows the same pattern, whether the subjects are transgender people undergoing cross-sex hormone therapy, aging people given testosterone supplementation, or prostate cancer patients treated with testosterone blockers.

Still, some T skeptics have found ingenious ways to avoid the inevitable conclusion. Here's an example from Jordan-Young and Karkazis's book *Testosterone*. They describe one of the cleanest, most influential studies ever conducted on testosterone and its effects on muscle in men, by the endocrinologist Shalender Bhasin and his colleagues:

> That classic study is the go-to citation for evidence that T builds muscle. But it's also a great study to look at to understand some of the limitations of that claim. First, *to find the effects of T on muscle, Bhasin and colleagues had to give huge doses of T*, six times more than had been studied in previous research on the effect of T on muscle. Second, even at these high levels of T, the significant increase in muscle size, and especially in strength, was mostly confined to the group that exercised regularly in addition to receiving T. *T alone didn't do much.*

There's only one problem: both of those italicized claims (italics are mine) are unwarranted. The Bhasin lab has produced several studies on testosterone's effect on muscle growth, and the one that the authors of *Testosterone* chose to discuss (from 1996) was designed specifically to evaluate the effects of very high doses of T on muscle growth, not to determine the level of T required to increase muscle. But even in that study, the researchers found that T increases muscle and strength in men who *don't* exercise. Here's a shocker: among the men who were on T (rather than placebo injections), those who exercised gained more muscle than those who didn't. Of course exercise plus testosterone adds up to more muscle. But "T alone" did plenty.

Let's now take a look at the Bhasin study that *was* designed to answer the question of how different doses of T—below, within, and above the normal male range—affect strength and muscle.

In that influential 2001 study, Bhasin and colleagues evaluated sixty-one young men (aged eighteen to thirty-five) who were given varying levels of T for twenty weeks. To ensure that they had full control over the subjects' levels of T, the researchers first blocked their

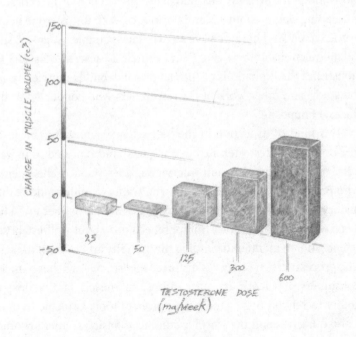

Results from Bhasin et al. 2001

natural T production. They used the highest-quality methods possible: the men were randomly assigned to one of five T-level conditions, and neither the researchers nor the subjects themselves were aware of the levels of T they were receiving. (This is called a "randomized, double-blind" study design.) *All of the subjects refrained from weight-bearing exercises* for the entire period of twenty weeks, and researchers evaluated the changes in the subjects' muscle volume and strength. Thigh muscle volume (results shown above), quad muscle volume, leg press strength, and fat-free mass changed in dose-response fashion, with increasing levels of T producing a greater increase. T significantly increased muscle and strength, within the healthy, normal range, without exercise. (See the results depicted above.) And crucially, these findings have been replicated by a second research group, using a similar protocol, and other studies have found similar results.

I wanted to take some time here to describe the methods and findings of Bhasin's studies because I'd like you to understand one of my motivations for writing this book: the prevalence of inaccurate or misleading claims about sex and hormones, with testosterone being a prime target. And here we have an excellent example. A "go-to" study on the relationship between T and muscle is described as one that "illustrates the limitations of the claim that T builds muscle." In fact, the study, and other work of the researchers who conducted it, does the exact opposite.

This kind of distortion of the best science, appearing not just in one book about testosterone but embraced and repeated in newspapers, magazine articles, and interviews, does science, the scientists who produce it, and the layperson who wants to understand reality a disservice. It makes my job as a science teacher harder because I have to take the time to correct misapprehensions about well-established science, not to mention defending the integrity and value of the scientific process. When science is distorted and studies are cherry-picked, consumers of science are left confused or misled, and we lose our ability to rely on one of the most powerful tools available to understand ourselves and the world around us. Bhasin's studies are among the most rigorous and well designed in the field of endocrinology.

Let's allow that kind of strong science to stand, and use the results to inform our thinking, discussions, and policies.

Testosterone builds muscle, men have more of it, and it gives them a strong advantage over women in sports.

ATHLETES WITH NATURAL T DIFFERENCES

Given the evidence so far, I think we would all predict that athletes competing in the women's division, who have DSDs like 5-alpha reductase deficiency, whose T levels are in the male range, would tend to outperform women within the normal T range. And in fact, that is exactly what we do find. By one estimate, people with these conditions are overrepresented in elite female sports by 140 times, compared to the prevalence in the general population.

The upper limit of T for women with PCOS, as reported in a comprehensive literature review, is still half that of men at the lowest end of the normal male range but can be as much as five times higher than the normal level for women. Although PCOS can be quite painful and may interfere with fertility, having it doesn't prevent women from participating in sports.

Given the high T levels in women with PCOS, it would be reasonable to predict that they would also be overrepresented in elite female sports—although their advantage should be slight, compared to athletes with 5-ARD (who have testes and much higher T levels). And again, this is what we find. In one study of ninety female Swedish Olympians, 37 percent had PCOS, approximately three times the general population rate for the same age group.

MORE T-SKEPTICAL RHETORIC

T's role in turning boys into bigger, faster, and stronger men is a staple of endocrinology textbooks. Nonetheless, the T skeptics often sound quite convincing.

Athletes with DSDs like 5-alpha reductase deficiency who compete in the women's category are frequently described in the popular press as "females with naturally high T levels" or "hyperandrogenic women," without mentioning their DSDs. This has the effect of making these athletes sound like the Olympic champion swimmer Michael Phelps, a man with a naturally—albeit freakishly—large arm span. If we want to find out the distribution of male arm spans, then there's no reason to exclude Phelps from the data. True, Phelps didn't win all those golds simply by training a lot—the genetic lottery gave him a big head start. But basically he's just a developmentally ordinary male.

Some athletes who compete and live as women also have XY sex chromosomes and testosterone-producing testicles. They are not simply women who are being forced to reduce their naturally high T, as is usually reported. Here's the opening paragraph from a 2019 article in the *New York Times*, reporting on the recent decision by the Court of Arbitration for Sport (CAS) on testosterone limits in athletes like Semenya:

> Female track athletes with naturally elevated levels of testosterone must decrease the hormone to participate in certain races at major competitions like the Olympics, the highest court in international sports said Wednesday in a landmark ruling amid the pitched debate over who can compete in women's events.

This is, of course, a sensitive issue because athletes like Semenya live and identify as female, and people want to respect their gender identity and privacy. However, the *Times* piece obscures the dramatic physical effects wrought by male-typical levels of T during crucial stages of development, and into adulthood. A more neutral opening paragraph would begin: "Track athletes who identify as female, who also have internal testes that produce levels of testosterone in the male range . . ."

When important policy decisions rest in significant part on people's sexed features, we need to get the facts straight. We don't do anyone any favors when we muddy up important discussions with misleading statements about biology. We don't need to twist scientific findings and terms in order to respect people's human rights, individuality, and worth. That's something we should do in any case.

Testosterone's potent effects on height, muscle mass, strength, and aerobic capacity via hemoglobin increases all confer obvious advantages in sports performance. Some people make the argument that it's not fair to focus on T levels because everybody has natural differences that affect performance. But apart from perhaps age and health status, only T draws such a clear and consistent line between large groups of people who differ in athletic ability.

SEX SEGREGATION IN SPORT

Caster Semenya is far from the first person whose body has presented a challenge to the binary classification of sex in sports. The IAAF and the International Olympic Committee (IOC) have a long history of failed attempts to balance the goals of fair competition with the rights of athletes who wish to compete as women but who have biological features that are more typical of males than females. These policies have involved women being subjected to compulsory tests, designed to verify that women have the "right" genitalia, genes, and chromosomes. Overall, these tests have been inaccurate and often humiliating to the athletes.

One low point was the mandatory "naked parades," which made their debut at the European Athletics Championships in Budapest in 1966. Like most other athletes subjected to them, Maren Seidler, a six-foot-two, powerfully built American shot-putter, to put it mildly, was not a fan: "They lined us up outside a room where there were three doctors sitting in a row behind desks. You had to go in and pull up your shirt and push down your pants. Then they just looked while you waited for them to confer and decide if you were OK. While I was in line I remember one of the sprinters, a tiny, skinny girl, came out shaking her head back and forth saying. 'Well, I failed, I didn't have enough up top. They say I can't run and I have to go home because I'm not "big" enough.'"

Sports are sex-segregated because of the advantages that male puberty brings, together with their continued maintenance by T during adulthood. Without sex segregation, people who didn't go through male puberty would effectively be shut out of elite competition. But segregation has its costs. If you're brought up as a female, are treated by everyone as a female, think of yourself as a female, and perhaps are also legally female (which is an option available to some people born male, depending on local laws), then of course you'd want to compete as a female.

And if you're a transgender woman, you might also feel that you should have the right to compete in women's sports. The dilemma arises because some athletes who quite reasonably want to compete in the women's category have also experienced the physical benefits of male puberty, and those benefits don't all disappear with the testosterone reduction that comes with a male-to-female gender transition. When evaluated after a year of T-suppressing medication, along with increasing estrogen, many of the sports-related benefits of high T decline significantly. For example, hemoglobin plummets to female levels. But bone size (including height, of course) doesn't budge, and much of T-induced bone strength is also retained. Experts and activists debate the question of just how much strength and muscle volume drop, but evidence shows that male-typical levels of muscle mass and strength are not completely lost, although the changes vary widely among individuals. In some trans women, no muscle at all is lost, and in others, levels drop considerably. One finding is clear, and that is that the muscle gain in female-to-male transgender people, who move from female to male levels of T, is significantly larger than the muscle lost in the other direction. These observations are based on studies of non-elite-athlete transgender populations, so effects of reduced T may be different in trained athletes.

However we resolve this dilemma about who can compete in women's sports, some people will feel understandably aggrieved. It's a sensitive issue that can cause stigmatization and discrimination, and I'm not sure how we should resolve it.

I'll leave the last word to the Court of Arbitration for Sport, which heard Caster Semenya's appeal against the IAAF's DSD regulations in 2019. Semenya and her representatives argued that the DSD regulations "lack a sound scientific basis; are unnecessary to ensure fair competition within the female classification; and are likely to cause grave, unjustified, and irreparable harm to affected female athletes." After hearing from many experts, including David Handelsman, the CAS dismissed her appeal. But they did say this:

> The Panel pays tribute to Ms. Semenya's grace and fortitude through-out this difficult process and expresses its gratitude for her dignified personal participation and the exemplary manner in which she has conducted herself throughout the proceeding.

Endocrinology by itself doesn't tell us whether athletes like Semenya should be allowed to compete in the female category. Whatever you think about that fraught issue, we can all agree that Semenya and other athletes with similar differences should be treated with respect and have their cases decided fairly and in the light of the facts.

ANTLERS AND AGGRESSION

WISDOM 11

In early October 2019, I found myself nestled into a recess on a rocky hillside on the Isle of Rum, off the west coast of Scotland. The spot gave me shelter from a biting wind, but it also allowed me to take in the view, which was spectacular. Stretching out ahead of me was Kilmore Glen, and rising up behind it and to my right were a series of undulating hills, blanketed by short grass and dotted with jagged rocks. To my left was Kilmory Bay, where I could see waves crashing ceaselessly against the craggy coastline.

I'd come to visit the Red Deer Project's research site on the island, a trip I'd fantasized about since my graduate school days. Getting there had involved many stages: a transatlantic flight; a five-hour train ride through the stunning scenery that is the Scottish highlands; an overnight in the small, picturesque town of Mallaig, on Scotland's west coast; and then a morning ferry to Rum—deer population about 1,000, human population 33. Ali, one of the project's two research assistants, met me at the dock, and we drove in the project's Land Rover toward the site on dirt roads through rolling sandstone hills covered with moss and short grasses. As we approached, I began to hear the deep, booming roars of the stags, who were competing for female deer, or hinds.

Now, from my hillside perch, I was looking down at a proud, majestic stag known to the researchers as Wisdom 11 (the stags are named after their mother and their birth year). He held his heavily antlered head high, supported by a long, thick neck from which a shaggy mane hung. The antlers branched up and out from each side of his head, curving in again at the top, each culminating in five sharp tips, reaching almost three feet in length. And sprouting straight up from his head, above his eyes, were two shorter branches, perfectly positioned to poke out the eyes of a rival. The front of his body looked quite bulky, but it was supported by relatively spindly legs. Big stags like Wisdom 11 are about seven feet long and weigh somewhere in the neighborhood of 450 pounds.

Wisdom 11 was surrounded by his harem of twenty-two slighter, smaller, antlerless hinds, which enhanced the impression of his size and power. Scattered about the hills and the long valley I could see five smaller harems, each with a single stag.

Early October on Rum is the height of the red deer mating season, known as the "rut." The place is abuzz with activity—and, as you might expect, testosterone. For those interested in how testosterone, sex, and aggression are related in wild animals, few places on earth offer a more spectacular illustration.

I had already read a lot about this population of deer—they've been studied on Rum since 1953, longer than almost any other population of wild vertebrates in the world. Over one hundred scientific articles and three books, containing groundbreaking contributions to evolutionary biology, have been published based on the research from this site. So I knew the data that described how the stags' age, size, and antlers all influenced the number of hinds they could acquire, and how many more offspring the winners would likely produce than the losers. In my classes, I shared this with my students as a potent illustration of how evolution can shape aggressive and sexual behavior in male animals.

But now the deer were right in front of me: the winners surrounded by their harems; and the losers, the solo stags (for now, anyway), wandering around on their own. I could see at least six such stags moving around the periphery of Wisdom 11's territory, covetously eyeing his harem. Some were relatively close to the harem in the valley, and others were at a safe remove on the hillside.

All stags are motivated to acquire territory and a harem, but each season only a few succeed. Many will manage to defend at least one hind for some of the time, but the success of stags like Wisdom 11 leaves most others spending the season as bachelors. The traits that predict status and the number of hinds in a harem are clear: stags need to be big, strong, and healthy, and neither too young nor too old (the sweet spot is between seven and ten years). Benefits of being the right age include experience and body size, but also the antlers, central to every fight, will have reached their peak size and strength. When two stags go at it, mingling antlers and pushing on their opponent, small or weak antlers are a liability.

Wisdom 11 with some of the hinds in his harem

Stags who are unlikely to get what they want via an "antler-to-antler" strategy must find other creative ways to try to overcome their celibate situation. On a few occasions when I left my perch to stretch my legs or get a better view of a scene below me, I was startled by a lone stag a few feet away. (I knew that the deer rarely attacked people, but that didn't stop my nervousness; he could have easily killed me if he had wanted.) Stags without harems would sometimes hide among the rocks in the hills, waiting until harem-holders like Wisdom 11 were distracted with head-of-harem responsibilities, including herding the hinds, mating, and chasing off other encroaching stags. When the opportunity arrived, one of the desperate stags would make a daring dash into the edge of Wisdom 11's territory, in the almost always vain hope of getting lucky with a hind. Stags can smell when hinds are in their short window of estrus, which can last for as little as two hours, and it draws those without harems of their own from near and far. The researchers call these stags who try to mate with hinds in another's harem "SFs," short for "sneaky fuckers." The lower-risk strategy can benefit younger stags in particular: they get the chance to practice being bold, but they avoid the high risk of injury that comes with a direct challenge to a dominant. And rarely, it can pay off with an actual mating.

Wisdom 11 stood tall atop a small mound and surveyed his surroundings. He tilted his head up so that his huge antlers hovered over his back, creating the impression of an even more formidable body. He let out a booming roar. Then another, then another, then another. I heard a sequence of roars in response, coming from one of the stags pacing around on the hills behind him—Tattler 06. Wisdom 11 turned to face Tattler 06, a solo stag who had run down from the hill and was now slowly approaching.

During the rut, an approach like this is never taken as a friendly greeting between two males. It's an unwelcome invasion of the broad boundaries of personal space, and a direct challenge. Both stags paused, stood tall and stiff, facing each other, and continued their roaring contest for another few minutes. At this point, many approaching stags figure out that their chances of victory are slim and do the prudent thing, which is to run away. But not Tattler 06. He was willing to escalate, and that meant that the two stags would take the next step on the road to battle: the parallel walk.

Like most people, red deer stags don't make a habit of recklessly jumping into a physical conflict. Fighting is risky and draining and is best reserved for when the rewards—usually hinds or increased dominance that might help to get hinds later—are worth the risk. Stags are built to fight hard, even to the death. But if they can get away with intimidation, they prefer to stop there and frighten off a rival by displaying their size, weaponry, and intentions. Rivals take a keen interest in these displays, which contain valuable information. For example, Rum researchers have shown that roars are "honest signals" of fighting ability: louder, deeper, and more frequent roars are given by larger, healthier males who win the most fights. The signals are considered honest because they can't be faked. Such roars can be produced only by those with the strength and body size, and enough energy in the moment to do so. And these are some of the same qualities that are needed to be victorious in battle.

If questions over who's boss can be worked out by evaluating fighting capacity rather than actually fighting, then both parties benefit. The loser lives to fight and mate another day, and the winner gets to maintain his harem. (Although if the loser has hinds of his own, the winner might take those, too.) Physical battles happen only when each party believes that victory is within his reach, or when desperation, usually provoked by the smell of estrous females, drives one of them to do something a little crazy, like darting into the middle of a star stag's territory to try to steal a mating.

Strength isn't everything. Skill and grit matter, too. As Tim Clutton-Brock, an evolutionary biologist at Cambridge University and long-term Rum researcher, describes:

> Many behavioral factors apparently affect an individual's chance of winning. Some stags fight skillfully, taking advantage of ground surface, slope, and their opponent's behavior, while others do not. Some are determined fighters, tenaciously holding their ground even after

they have been pushed back several times by a larger stag, while others give up easily.

These intangible qualities of a stag's opponent are harder to assess in advance, and information about one's chances of victory may not be available until the battle commences.

Wisdom 11 and Tattler 06 moved slowly toward each other, holding their heads high, showing off their strong necks and antlers, stepping stiffly and slowly, maximizing the impression of their ability to inflict harm. Then the two big males turned and walked side by side, about five meters apart, looking straight ahead. The stiff-legged, tense walking continued for a few minutes, back and forth, back and forth. Suddenly Tattler 06 turned to face Wisdom 11. He bent his weaponized head down in a formal invitation to battle, and Wisdom 11 turned and accepted, also lowering his antlers. I heard a knocking of hard bone as the antlers of the two animals intermingled and watched each stag struggle to push against the other, trying to throw him off balance and wrestle him to the ground. If one goes down, that gives the victorious stag an opportunity to plunge an antler tip into his opponent's neck or flanks, which he will not hesitate to do.

Fighting stags

Then, just as suddenly as it started, the fight stopped. The two disengaged and resumed parallel walking, as though just a few seconds before they hadn't been trying to poke each other's eyes out. That lasted for a minute or so, then Tattler 06 lowered his head and invited his rival to resume battle. It all struck me as rather gentlemanly. No cheating, no funny stuff. They were following all the rituals on the road to battle I'd read and taught about for many years. The next round didn't last long, less than a minute. Soon Tattler 06 ran away, back to the safety of the hills, and Wisdom 11 returned to his harem.

I saw a dozen or so such battles during my four-day stay on the island—some shorter, some longer, and one with an antler jab to the face that caused a bloody injury. I even saw Wisdom 11 challenge his next-door neighbor and harem holder Glariola 09, who had only four hinds. After a short fight, Glariola 09 ran off, leaving his hinds to Wisdom 11, who added them to his already impressive collection.

While all this stag action captured most of my attention, the contrasting behavior of the hinds was also remarkable. Each one was, like the stags, focused on what she needed to do to reproduce. For a hind, that meant occasionally mating, but mostly eating, resting, and trying to avoid harassment from the harem holder. He doesn't like it when she strays too far or doesn't mate on demand, so she complies, in order to avoid being barked at, chased, or sometimes kicked. (She isn't stuck with him, though. If she really doesn't like it, she can run away to another harem or strike out on her own.) If all went well, and she stayed healthy and got the nutrients and calories she needed, her new calf would be born the following spring.

I think you can guess which hormone is responsible for the striking features of the stags that differentiate them from the hinds, including the stags' inclinations to fight and mate, often in rapid succession. The red deer of Rum are the perfect introduction to how and why testosterone promotes competition that sometimes culminates in violence between adult males.

CHILLING OUT

For most of the year, the competitive stakes are low for Wisdom 11, Tattler 06, Glariola 09, and the rest of the stags: the hinds can't get pregnant and aren't sexually attractive. They live separately from the stags, in groups of mostly female relatives, tending to their calves and grazing. In their bachelor herds, the stags still know who's boss, and when status does need to be settled, conflicts are usually resolved without resorting to an all-out brawl. There's no roaring or parallel walking, and no clash of antlers, which, in any case, fall off after the rut.

Hinds live a relatively calm and stable existence year-round. They periodically get into skirmishes, but nothing vicious. They'll chase, nudge with their nose, and, more rarely, kick or rear back on their hind legs and use their front legs to duke it out. (This "boxing" is also how the stags fight outside of the rut.) And being top hind can pay off with increased access to the best feeding areas, leading to a better physical condition, which translates into more surviving calves over a lifetime. But since there's no equivalent of what the stags are fighting for each season, which is something akin to a reproductive jackpot, hinds don't need to be so aggressive. They need to stay well fed and healthy. A successful hind might bear a calf every year with a maximum of about fourteen over a lifetime, but a successful stag, like Wisdom 11, could father seven in a year for a grand total of up to thirty. Last year he sired about fifteen calves, which was a whopping 25 percent of the total number of calves born that season, and a world record for the stags of Rum. And this was in a group of nearly ninety adult males.

Stags and hinds both need lots of energy to fuel different reproductive strategies, and the sex hormones help to direct that energy as needed.

A busy, dominant harem holder doesn't have time for luxuries like eating during the rut, when he spends only about 5 percent of his time feeding. So in the off-season, he spends most of his time eating and resting, as do the females all year round. He also has to beef up and store fat to supply the energy he will need to fuel his focus on competition and mating. He must also stay injury free. But he won't be able to accomplish these goals if he's a hot-tempered hooligan, so it's to everyone's benefit if they can learn to get along.

One physiological change promotes this stag serenity: after the rut, the testes slowly shut down, and blood T drops to very low levels. It's like a temporary castration. In the absence of any reproductive prospects, sperm and T aren't needed. And the stags chill.

ANTLERING UP

As the length of day shortens toward the end of summer and fall approaches again, the stags' testicles begin to emerge from their slumber. They grow in size and triple in weight, and ramp up T production to maximum levels. And the hormone's earlier actions on the body and brain ensure that seasonal elevations in T can get the job done without having to start from scratch. Testosterone lays the foundation for aggressive competition before the deer even become reproductively viable. Prenatally, T masculinizes the male's reproductive system. Then during puberty, male-typical T levels promote longer and stronger bones and enhanced muscle mass—a big stag can weigh almost twice as much as an average hind. T also elongates a stag's larynx during puberty. A longer vocal tract produces a deeper voice; like those men with voices that make them sound like they should be in charge, a stag's deeper voice intimidates rivals and attracts females.

Testosterone levels heading into the rut are vastly higher than they need to be to support sperm production, which will continue as usual even when T levels are relatively low. So what's all that "extra" T for? It's not about sex, it's about war. All that T helps to make the males more intimidating and better fighters.

The stag's antlers were cast off at the very end of the rut, in the previous spring. (See the graphic below for seasonal changes in T, aggression, and antler growth.) They begin growing right away, but until the end of the summer they are not very strong and are covered

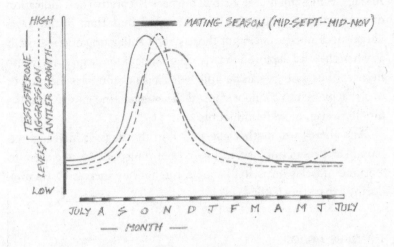

Stag T, aggression, and antler changes across the seasons

with a soft, velvet coating. The velvet provides a blood supply to the growing bones, enriching them with growth factors and other nutrients, enabling them to grow up to nearly two centimeters per day. (Antler velvet is a hot commodity online, with sellers claiming that it reduces stress and erectile dysfunction and increases libido, strength, and even T production.) A stag at this stage is like a knight with a plastic sword glued in a leather scabbard.

When T rises in August and September, it ensures that a stag can make good use of the huge, bony structure on his head. As in humans, high T increases the calcification of bone, strengthening it. In stags, this calcification happens preferentially in the antlers. And as a result, antler bone is about three times as strong as the bone in the rest of the stag's body. Stags with higher T have stronger antlers that are more resistant to breakage during battle, enabling them to more readily ascend the status hierarchy. That high T also cuts off the blood supply from the velvet, causing it to be shed and revealing the antlers' sharp tips. No more messing around.

High T further prepares a stag in anticipation of the rut. It promotes the growth of muscles around a stag's neck, doubling its girth.

In stags, as in humans, the girth of the neck is a pretty good indication of one's fighting ability—but it's even more important in the stags. Larger neck muscles help put those strong antlers to good use, such as when they are deployed to twist an opponent to the ground, where more serious damage can be inflicted. T also encourages the growth of a shaggy mane around a stag's neck, exaggerating its size, making him look even more intimidating.

And there's yet another effect of T in the stags: it increases red blood cell production, increasing oxygen transport. More oxygen to working muscles helps stags stay in the fight by increasing stamina during exhausting battles.

T AND BEHAVIOR

High testosterone clearly helps red deer stags fight. And of course, without T, there would be no DNA-carrying sperm to transmit their genes into subsequent generations of red deer. But what about the stags' behavioral metamorphosis, from friendly to fierce? Was that caused by testosterone acting directly on a stag's brain, altering neural circuits and disposing him toward aggressive behavior? Well, maybe, but that's just one hypothesis. As we all know, correlation or association doesn't equal causation. And even if T does cause aggressive behavior, it might do that without directly altering the brain.

One rival hypothesis is that the stags become more aggressive simply because T endows them with mighty antlers and muscles, which then influence social relationships, which, in turn, impact behavior. Or T could increase sexual interest in the hinds, which then inspires the stags to compete. Or maybe the aggressive behavior is not explained by T at all but by something else. The causal relationship could even be the reverse—aggressive behavior causes high T.

How could we find out? Let's take another example familiar to many parents: sugar and children's activity level. My son, Griffin, always looks forward to Halloween, but after he's eaten so much candy, there's just no getting him to bed on time. What was the cause of this bonkers bedtime behavior? One hypothesis is that the sugar he and the other little candy

ANTLERS AND AGGRESSION   141

consumers ingested is causing them to run around, perhaps because the body needs to spend that "extra" energy. Many of us have seen with our own two eyes that kids get rambunctious after sweets, so we might reasonably assume that the sugar caused the silliness.

But what if Griffin's high energy wasn't caused by sugar but was just the excitement of Halloween? He had spent a fun night with friends, dressed up in a space alien costume, and was looking forward to eating mounds of candy every day for the foreseeable future. On this hypothesis, sugar is only correlated with excited behavior and is not the cause of it.

How could we determine if the sugar caused the craziness or if it was just associated with the real cause? We could consider our observations in the context of other information—is there a known mechanism through which elevated blood sugar could cause an increase in activity level? Do nonhuman animals also go a little nuts after too much sugar? Is there independent evidence that social environment influences children's activity levels?

We could also do a kind of "remove and replace" experiment, like Berthold did with rooster testes, as I described in chapter 2. But this time, we remove and replace the sugar in the candy. Keeping the environment as consistent as possible, Griffin could eat sugarless M&M's (that by some miracle tasted just like the real thing) before bed for one week, followed by a week of ordinary (sugar-sweetened) M&M's. The sugar hypothesis would be supported if he had a significantly harder time relaxing before bed the week he ate the sugary M&M's. But if he had no trouble at all relaxing and getting into bed every night for two weeks—ha!—that would leave the door open for the "party" hypothesis.

You get the picture. Careful experiments are needed to show what causes what, and consistency with other well-established theories is also important. (Experiments along these lines have actually been done for the sugar hypothesis, and it turns out that it is a myth.)

Hormone-behavior relationships are almost never straightforward, especially when they exist in complex social and ecological environments, like in most wild animals. But when it comes to physical aggression, "remove and replace" experiments across multiple species, in the wild and in captivity, show that T belongs front and center. Such experiments were carried out on the stags of Rum in the 1970s, some the first of their kind ever to be performed on wild animals.

To determine the role of T in aggressive behavior of the red deer, researchers castrated three Rum stags at different times of the year, in and out of the breeding season. Deprived of testosterone, their antlers were soon cast. New antlers began to grow, but they were malformed, lacked branching, and remained covered with velvet when it would have normally been shed. Their neck muscles remained in their smaller, summer condition, and the shaggy mane around it failed to grow in at the usual time prior to the rut. When the intact stags shed their velvet and came into "hard horn," the castrated stags with their soft small horns couldn't compete even if they had wanted to. Their aggression, and status, plummeted. But they did not seem to mind. The once sexually potent scents from estrous hinds failed to excite them, and the castrated stags didn't even bother to roar or visit their usual rutting areas.

Castrating stags caused aggression to drop, but the researchers couldn't be sure whether the change in behavior was due to the absence of testosterone acting on the brain or to another, less direct effect of T. Perhaps what matters is the absence of T acting on the horns. The stags might have been more placid because they, or their peers, realized they weren't worthy competitors.

To answer that question, the researchers later replaced the T. They implanted the castrated stags with levels of testosterone that would be typical for the rut, using long-acting, slow-release capsules. T was increased in this way at two times of the year, inside and outside of the normal rutting period, to determine the effects on sexual and aggressive behavior. When T was returned to those antler-less, castrated stags, their interest in the hinds also returned—the implanted stags sought out the hinds and attempted to procure a harem. But this sex effect occurred only during the rut, when the hinds were fertile. When the hinds were not fertile, outside of the rut period, the high T had no effect on the stags' sexual interest. Presumably this was because the other environmental cues that help to stimulate sexual behavior weren't present, like the scent of hind fertility and an alteration in day length that signals the rut.

Those newly high-T stags may not have cared about sex in the off-season, but they still cared about dominance. And it didn't matter what time of the year it was—in or out of the rut. With the T implants, the previously docile stags began to pick fights with nearby males, even when their antlers had not yet grown back and were still covered with soft velvet.

These results show that testosterone is necessary but not sufficient for the expression of sexual behavior—the proper environmental stimulus, whether fertile females or something associated with their fertility, must also be present. And the results also show that T's effects on male-male aggression are most likely not due to something associated with T, like having big, sharp antlers, receiving aggressive behavior from other males, or even the presence of fertile females and a high libido. Experiments of this sort never amount to conclusive proof, but there aren't any serious rival explanations. The evidence we do have, from red deer and many other animals, strongly suggests that T promotes aggression in these animals by acting on the brain.

Although sex and aggression often go together, the capacity of T to mediate the two traits separately makes evolutionary sense. In many male animals, including seasonal breeders (like the red deer) and nonseasonal breeders (like chimpanzees), dominance relationships are established and renegotiated when no fertile females are present. These relationships affect the resources males can acquire, like territory, that will later affect their ability to find mates. In other words, aggression pays off even when mates are not directly at stake. Changes in T can help males plan for their reproductive future.

SELECTION AND SEX

Some stags will never father any offspring. From an evolutionary point of view, they might as well be dead, since their genes will end with their lives. This means that stags who monopolize hinds and shut out other stags, even at great risk of injury, have an evolutionary advantage. Those with genes that promote slightly larger, more muscular bodies or sharper antlers will tend to have more offspring than their rivals, and those who are just a bit bolder, and able to maim or blind another stag, will also tend to benefit reproductively. That process helps to explain how, over many generations, the male and female forms of this species have become distinct.

Red deer are a compelling example of T's effects in male animals partly because they are seasonal breeders. In concert with seasonal fluctuations in the social and physical environment, T's highs and lows induce clearly observable and dramatic changes, transitioning stags from spermless, weaponless, relatively peaceful animals to highly sexed, aggressive, and dangerous ones. But red deer share many of the sexual asymmetries that are common across mammals—most relevantly, the males are able to produce offspring at a faster rate than females.

This asymmetry begins with the size and quantity of eggs (large and limited) versus sperm (small, plentiful, and continuously produced) and continues with the nature of the female versus male mammalian bodies. Females must use their bodies to host and feed their developing offspring, during which time they are unable to produce another offspring. But most male mammals contribute only DNA to each offspring and are free to invest the "extra" time and energy into the pursuit of additional mates. These differences lead to a predictable, sex-based behavioral pattern: males tend to prioritize competing for mates, and females prioritize acquiring the resources they need for health and survival and selecting fit males with whom to mate.

The British naturalist Charles Darwin was the first to describe this pattern of sex differences, and the explanation for it, in *On the Origin of Species*, published in 1859. "This form of selection," he wrote, "depends, not on a struggle for existence in relation to other organic beings or to external conditions, but on a struggle between the individuals of one sex, generally the males, for the possession of the other sex." As mentioned in chapter 2, Darwin called this *sexual selection*. Sexual selection does not favor traits that help in the "struggle for existence," like a shaggy coat to protect from the cold or camouflage to fool predators. Rather, it favors traits that increase the animal's ability to acquire mates. Antlers help make babies because they help stags get hinds.

Competition between one sex (as Darwin says, usually males) for mating opportunities is only one type of sexual selection. The other involves one sex (usually females) choosing among members of the other sex, a familiar phenomenon in birds. As Darwin put it, "birds of paradise, and some others, congregate, and successive males display with the most elaborate care, and show off in the best manner, their gorgeous plumage; they likewise perform strange antics before the females, which, standing by as spectators, at last choose the most attractive partner."

The quintessence of this kind of sexual selection by "mate choice" is the peacock, with his train of long, brilliantly colored and decorated feathers. The peahen's backside, by contrast, looks stunted and dull. Darwin puzzled over this kind of sexual dimorphism for a long time, and famously complained in an 1860 letter to his friend, the Harvard botanist Asa Gray, "The sight of a feather in a peacock's tail, whenever I gaze at it, it makes me sick!"

Darwin explained the ornamentation of male birds in his second masterpiece, *The Descent of Man, and Selection in Relation to Sex*, published in 1871:

> There is not much difficulty in understanding how male birds have gradually acquired their ornamental characters. All animals present individual differences, and as man can modify his domesticated birds by selecting the individuals which appear to him the most beautiful, so the habitual or even occasional preference by the female of the more attractive males would almost certainly lead to their modification; and such modifications might in the course of time be augmented to almost any extent, compatible with the existence of the species.

Selection by mate choice is more complicated than Darwin realized, but his basic idea was right. Mate choice explains some aspects of sexual dimorphism in many species, not just in birds but also in amphibians, fish, reptiles, and primates. When females actively choose certain males for mates, whether it be the beautiful, bold, melodious, mean, or fragrant, her "decisions" are likely to be forceful drivers of the evolution of his secondary sex characteristics—that is, those that differ by sex, emerge during puberty, and are not directly involved in the physical act of reproduction.

FEMALE AGGRESSION

Evolution can produce high rates and intensity of aggression in female animals, particularly when they need to compete directly for resources like food, nesting sites, or males. These reproductive resources can be provided indirectly by a mate (who can also supply good genes), or in some cases, they can be increased by eliminating the ability of the competition to reproduce. A striking example is the female naked mole rat, which lives in underground colonies in the African desert. It is a small hairless rodent with wrinkly pink skin, and it looks like a penis with teeth. The females are dominant to the males, and an aggressive female can push around and harass other females so much that the stress causes their ovarian function to shut down, rendering them infertile. The queen naked mole rat then has her preferred males for herself.

And then there's the infamous female spotted hyena, highly aggressive, giving even seasoned experts difficulty trying to tell her apart from the male. (Her clitoris, through which she pees, has sex, and gives birth, looks exactly like a penis. She even has a convincing-looking fake scrotum.)

One more example is the female meerkat, who, when dominant, monopolizes breeding opportunities within a group for up to ten years. Becoming top meerkat pays off, and she'll fight hard for it—among other nasty tactics, by killing off the babies of her female rivals.

In all these cases, the endocrine modulation of female aggression is not well understood, and there's no clear relationship to testosterone as there is in males. But there is some evidence that prenatal T is important in some of these animals. For example, in the meerkat and hyena, dominant pregnant females have higher T levels than pregnant subordinates, and babies born to those high-ranking mothers end up more aggressive than the offspring of the subordinates. High T may be acting to shape the brains of these female fetuses to enhance their adult aggression.

For the most part, when females benefit from being intensely aggressive, T is not the hormone recruited to direct aggression toward the proper targets at the right time. Females have other tricks up their sleeve when aggression pays off reproductively, and they do not need to rely on the hormonal mechanisms of the other sex.

Females can be aggressive when their reproductive success is at stake, such as when their children are threatened or when competition over resources, mates, or the status required to attain those things is challenged. But overall, relative to male animals, females benefit more from being safe and cautious and living a long, healthy life. And lower levels of aggression help to make that possible.

COORDINATION AND COMMUNICATION

Sexual selection clearly acts on the body, but it also acts on behavior. This makes sense: it would be bizarre to endow animals with special weaponry like antlers or a spectacular tail if the animals had no inclination to use them to threaten others or attract mates. Evolution doesn't like to waste energy, and genes for expensive and useless traits tend to get weeded out of populations. Stags are motivated to break the leg or poke out the eye of a competitor if the opportunity arises, especially one who would be forced to relinquish his hinds because of his injury. And the inclination to fight for and mate with those hinds comes at just the right time, when they are able to conceive.

Strong and ruthless fighters with a zeal for sex sire more offspring, which then carry their fathers' genes for a high libido and bellicosity. These will be upregulated by T in the next generation of males. Conveniently, T also solves the problem of what to do with those genes in females: low T keeps them largely packed away unopened.

If you have weapons for only part of the time, you need to make sure that you're pushing other males around only when you've actually got those weapons. Otherwise, you'll be like a robber trying to conduct a heist with a banana—something that happens from time to time but never seems to end well. And if you don't have weapons to fight with or the opportunity to use them, sperm might not do you much good, either. Coordination and communication are testosterone's job, and it is an efficient performer. T is sexual selection's solution to the problem of how to match reproductive anatomy and physiology with the behaviors needed to put them to good use.

LOSER LIZARDS AND DEVOTED DADS

Mountain spiny lizards, found only in the mountains of southeastern Arizona, are polygynous seasonal breeders. They're small antlerless reptiles, but reproductively, they're remarkably similar to the red deer. Outside of the breeding season, the males live in close quarters, or "aggregations," where they regularly get pushed around and stepped on by their fellow lizards. But they're cool with it.

The male lizards have three different levels of territorial aggression: almost none in winter and early spring, which are outside of the breeding season; low in the summer when they are setting up territories; and high in the fall breeding season. During winter and spring, they need to store energy and to stay out of trouble (i.e., avoid needless conflicts) so that they can be in top fighting condition for the breeding season. To maintain the peace and increase the fat, their T levels bottom out in the winter. But come summer, rancor sets in and the males return to their breeding grounds to begin their annual competition for status and territory and, eventually, mates.

The females won't show up until the fall. The males save their fiercest aggression for when they arrive and they have something more valuable to fight over. But for now, in the summer, rather than roaring and strutting, these guys show off by doing a lizard brand of push-up and a few other flashy moves like head bobbing and shuddering, which, evidently, can be intimidating. Ignore a lizard push-up at your peril! If a male neighbor is indifferent to a resident's displays, he can expect to be on the receiving end of a charge or even a bite. To support this level of aggression, T is at the Goldilocks level: ten times higher than the winter basal level. Not too high, nor too low, but just right.

But what's just right about having T in the middle zone? What's wrong with having high T? If that is so great for male reproduction, why not crank it up beyond the Goldilocks level in advance of the females' arrival? Then a lizard could do more and fiercer push-ups, get more territory, and, eventually, more females and babies. Then again, why not just turn up the T dial for the entire year? Wouldn't a lizard or stag with perpetually high T have a head start on crushing the opposition well before the game even starts?

Scientists have wondered the same thing, and in the late 1980s they began to conduct experiments to find out the answers. Predictably, as the researchers learned, if you castrate lizards and remove T during the fall breeding season, the territorial behavior and sexual interest in the females decline. And if you take one of those normal, territorial, T-in-the-low-to-middle-zone *summer* lizards, and crank up his T to maximum levels that aren't normally reached until the fall breeding season—one hundred times the basal level of T—then territoriality and aggressive displays increase. (But not to maximum levels when there are no cues from fertile females.)

Stop the presses! Science shows that T increases bellicosity and horniness! But the researchers still hadn't answered the question about why males don't maintain higher T levels in order to get an edge on the competition.

At the end of the summer, they compared the lizards whose T had been increased to the "control" lizards who got empty implants—those with the normal Goldilocks, middle-zone T levels. Eighty percent of the Goldilocks lizards were still alive. They did what these lizards usually do in the summer. They spent about three hours a day outside of their hideouts, basking in the sun, chowing down on their favorite insects, and defending their territories. Meanwhile, the jacked-up lizards more than doubled the amount of time they spent out and about, patrolling their territories, attacking other lizards, and showing off. More time spending energy, and less time resting and eating.

Many of these high-T lizards did manage to increase the size of their territories. But they failed miserably in their preparations for the fall mating season, when those large territories would have paid off. Relative to the Goldilocks controls, the high-T males were now quite thin, or worse, dead. Half of them had met an untimely demise. They had come out of the gate too fast and spent their precious energy early and unwisely. The Goldilocks guys were rested, fattened up, and prepared for the arrival of the females. They got their more sensible genes into the next generations. Who's the loser now?

DEVOTED DADS

There's another way to be a high-T loser, and it was discovered by the aptly named John Wingfield, a British evolutionary biologist and avian enthusiast. As a young boy growing up in rural England in the 1950s and '60s, Wingfield delighted in nature and developed a driving curiosity about the seasonal changes in the behavior of local birds. His work has been among the most influential in all of behavioral endocrinology, much of it focusing on the hormonal regulation of the reproductive behavior of song sparrows living in the Eastern United States.

Like most birds, these medium-size brown-and-white ones with melodious songs are seasonal breeders. Their colors, singing, sex lives, and competitive tendencies change as the dark cold of winter gives way to spring. As with other seasonal breeders such as red deer and spiny lizards, when sperm aren't needed and the testes are mostly off-line, relations among the males are relatively smooth. But when temperatures rise and females become fertile, males compete for prime real estate (mostly by singing away, but sometimes backed up by physical aggression) to try to attract the females with the best reproductive prospects. Usually a male will settle down with one female for the entire season in the territory he defends, and together the pair will strive to raise a few broods.

Throughout the breeding season the female lays several clutches of eggs and cycles in and out of fertility along the way. When she is fertile, she is sexually attractive not just to her mate but also to local males, which results in a fair amount of cheating—about a quarter of song sparrow chicks are fathered by neighboring males! So her mate must "mate-guard" her, ensuring that other males keep their distance. He must also feed and keep his partner and chicks safe from predators (she mainly stays at the nest). When male-male competition is most intense, at the start of the season, T levels are at their highest. But when the initial commotion has ended and the birds have paired up, testosterone drops to the Goldilocks level—just enough to maintain what's needed for mating and raising the young.

A simplified depiction of this relationship between T and song sparrow breeding behavior in males is shown in the image below:

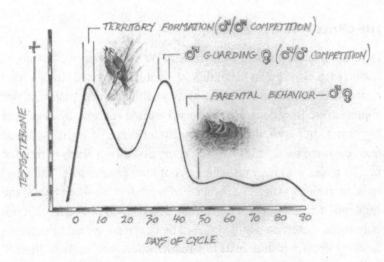

Variation in song sparrow reproductive behavior and T levels

Wingfield showed that these behavioral changes aren't just associated with changes in T but are regulated by them. You already know what happens when roosters, red deer, and lizards are castrated—sex and aggression plummet. The same thing happens, Wingfield found, in breeding male song sparrows. No T, no reproductive behavior. But what would happen if T were instead jacked up—this time, in males who are doing their daddy duties, when T levels would normally be at the Goldilocks level? Would the birds suffer as did the lizards, whose price for their T-enlarged ego was starving, sometimes to death? Not exactly. There *was* a price to pay—but this time it was the chicks who had to pay it.

Wingfield increased T in a group of males who were busy being dads, spending the day searching for tasty morsels like beetles, seeds, or worms and bringing them back to the nest. With elevated testosterone, other activities took on a greater appeal. Instead of devoting time and resources to their children, high-T dads went out singing at all hours along the perimeter of their territories, telling the neighbors to screw off and trying to score new females. The high-T dads neglected their families, and their chicks were more likely to die of starvation.

THE CHALLENGE HYPOTHESIS

In yet another song sparrow experiment, Wingfield plopped "intruder" males (each in a cage) in the middle of territories that had already been established. The caged birds sang away, and the resident males did not appreciate it. They defended their hard-earned territory by singing at high rates, and even attacking the intruder's cage. Wingfield and his team captured these irate birds and compared their T levels to those of nearby males who were just going about their regular lives. The territorial intrusions—these "challenges" to reproduction—had caused the residents' T to shoot up to maximum levels. It was now clear that not only could T increase aggression, but that aggression could elevate T.

Song sparrows, like males of many other species, keep their T levels as low as possible most of the time to avoid adverse effects on their health, survival, and reproductive success. This is made possible by stable social systems and status signaling, which diffuse everyday aggression, obviating the need to keep T high. But during times of social instability, when males must compete intensely over mating rights or the status or resources needed to secure them, T rises to the challenge. In short, T levels fluctuate depending on whether a male needs to be ready to breed, care for his family, or fight off rivals. In a 1990 paper, Wingfield called this the *Challenge Hypothesis*.

The Challenge Hypothesis makes sense of a diversity of findings that link T to aggression across vertebrates. Male T levels are exquisitely responsive to signals from the physical environment, like day length and temperature in seasonal breeders, that are reliable cues to ramp up (or down) reproductive physiology and behavior. But it is the cues from the social environment, like threats from rival males, sexy females, or the squawking of hungry young, that tell males when paying the cost of high T is worth it, or when it must be reduced and aggression saved for when it makes more reproductive sense. The ups *and* downs of T are adaptive: high T is not always better than low T, and depending on the circumstances, it can be fatal. Although Wingfield's initial formulation of his Challenge Hypothesis has been revised and updated to accommodate new findings, hundreds of studies have confirmed its basic framework for understanding testosterone-behavior relationships in seasonal and nonseasonal breeders, in monogamous and polygamous animals, from birds to fish to mammals—even insects. The mechanisms that underlie the Challenge Hypothesis exemplify the elegance of sexual selection.

As the reproductive endocrinologist (and one of my dissertation advisers) Peter Ellison has put it, "Male reproductive physiology can be crudely characterized as a system for turning energy into mating opportunities. It is in the management of this system that the functional significance of testosterone variation seems to lie."

There's little doubt that's true of nonhuman mammals. But what about us?

VIOLENT MEN

SIT THE FUCK DOWN

The writer Daemon Fairless and his wife, Lyana, had spent a pleasant New Year's Eve afternoon together, skating on an outdoor rink in Toronto, Canada. They were on the subway, looking forward to celebrating the New Year at the home of Lyana's parents. The festive atmosphere inside the train showed that their fellow passengers had already begun to celebrate. But, as the ride progressed, a group of rambunctious men increasingly drew the attention of the other, less inebriated passengers.

Fairless watched in growing irritation and incredulity as one inebriated man in his early twenties made a show of attempting to pry open the doors and stick his head out as the train sped through a tunnel. Voices in the car dropped off, as the other riders nervously eyed the antics. "A subhuman moron," Fairless thought. The loud foul language and the growing unease was becoming too much to tolerate. Fairless assessed his chances of taking control of the situation. He noted his advantages. Although the other man was big and perhaps fifteen years younger, Fairless was bigger—six feet, three inches tall and weighing about two hundred pounds. And, as far as he could tell, he was in better physical shape, as well as being sober. He determined that he could take him, if it came to that. Fairless recounted how things quickly got out of hand in his 2018 book, *Mad Blood Stirring: The Inner Lives of Violent Men*:

There's . . . a building surge on the horizon, moving toward me quickly, an impending wave, a deep, seductive swell. He's a piece of trash. There's a lapping coolness between my temples. People all around me are scared, nervous. They feel threatened. Not me. It's different in me, it's a growing itch, a form of lust. A predatory hard-on. I want him on his knees. Subjugated. Scared. I'm out of my seat. I'm standing beside him.

"You need to sit the fuck down," I tell the guy. My face feels tense. I'm baring my teeth, grimacing weirdly.

He looks up, genuinely surprised. He sizes me up and then cocks his head. "Who the fuck're you?" he yells. His breath is hot. . . .

I lean in, almost whispering in his ear. "I'm the guy," I hiss, "who can *make* you sit down." . . .

I can't remember what he says—something with *bitch* in it. His fist is raised over his head. He is way the fuck too close to Lyana.

I'm standing. We're chest to chest.

"You keep talking," I say, "but you haven't thrown a fucking punch. You're a fucking pussy. Throw a fucking punch. Otherwise sit the fuck down." . . .

The wave crashes. Everything is suddenly clear and simple. The solution is obvious. I'm relieved by the simplicity of it.

I bring my forehead down like a hammer. I aim for the bridge of his nose, but evidently he turns his head. I see stars—literally, cartoon stars. The guy reels back, but he's still standing.

Attack! The voice in my head is screaming. *Attack! Attack!*

The fight continued with the other man pouncing on top of Fairless. Fellow riders managed to pull them apart, but not before Fairless tried to poke his opponent's eye out in an effort to get him off. The drunken man was eventually handcuffed and taken away by the police, and nobody sustained serious injuries.

A dramatic story, but also easy to believe: two men literally butting heads (albeit without built-in weapons, like the red deer have for such occasions). But there is one adjustment that would make the interaction hard to imagine: change the sex of the protagonists. Not that women aren't aggressive. We are certainly capable of anger and inflicting pain, just as much as men. But when it comes to how we express that anger and hurt others, men and women are worlds apart.

TOXIC MASCULINITY

It's popular today to blame high rates of male aggression on the patriarchy and its social codes. Those codes, the theory goes, prompt men and women alike to teach boys, but not girls, that emotions and weakness are bad and that stoicism and aggression are good.

Here, for example, is how the American Psychological Association put the theory in "Harmful Masculinity and Violence," a newsletter that has circulated widely since its release in 2018. "Primary gender-role socialization aims to uphold patriarchal codes," the authors wrote, "by requiring men to achieve dominant and aggressive behaviors. The concept of gender roles is not cast as a biological phenomenon, but rather a psychological and socially constructed set of ideas that are malleable."

Matthew Gutmann, professor of anthropology at Brown University and the author of the 2019 book *Are Men Animals?*, agrees. He says that new research that is "just now reaching the general public" shows there is "little relation between T and aggression (except at very high or very low levels)." This, along with his interpretations of other scientific literature, has convinced him that biology, and testosterone in particular, is not where explanations of male violence are to be found: "If you believe that T says something meaningful about how men act and think, you're fooling yourself. Men behave the way they do because culture allows it, not because biology requires it."

Behavior is always a product of interactions between an animal's external environment and its biology, including its genes. And—to repeat one of the main points of this book—testosterone's primary job is to coordinate male sexual anatomy, physiology, and behavior in the service of reproduction. For many male animals that must compete for mates, like the red deer of Rum, one of the behaviors that most directly supports reproduction is aggression.

T's central role in male violence is well established for many non-human animals. Could men really be exceptions?

THE PURPOSE OF AGGRESSION

Aggression, defined broadly, is behavior intended to harm (or at least intimidate) another. It's a fact of life. Animals do what they need to do to survive and be reproductively successful. They need to eat, find mates, avoid being eaten, and ensure that enough of their offspring are able to reproduce themselves. Sometimes animals have nonaggressive strategies for achieving these ends—they have a sensitive nose for food; they make themselves look attractive to the other sex; they hide from predators; they produce thousands of young, a few of which will beat the odds and have their own offspring. Other strategies involve physical aggression: fighting off hungry and romantic rivals and predators that threaten either parent or offspring. Aggression is a strategy used by females as well as males throughout the animal kingdom.

But when the two sexes face different challenges to succeeding reproductively, the solutions for each sex will be also be different. For males more than females, reproductive success is limited by access to mates. That means that primarily in males, the solution, sculpted through the forces of sexual selection, is to develop traits that enhance fighting ability, like weaponry and the motivation to fight rivals.

IT'S NOT ALL ABOUT MEN

Although the stereotype of men as the more physically aggressive sex is backed up by a wealth of data, it would be a mistake to think of women as incapable of promoting—and sometimes of carrying out—extreme acts of violence. In 1994, during the genocide in the East African country of Rwanda, during which at least half a million people were killed, Pauline Nyiramasuhuko was the minister for Family Welfare and the Advancement of Women. She was later convicted on charges of genocidal rape. One witness recounted that right after Nyiramasuhuko ordered militia members to burn seventy women and girls using gasoline she had in her car, she said, "Why don't you rape them before you kill them?"

It's true that men are more physically aggressive than women overall, but women can also be physically aggressive. Intimate-partner violence—violence or physical aggression between a current or previous partner or spouse—is distressingly common and severely underreported. Research on sex differences in intimate-partner violence is controversial, methods are variable, and in many parts of the world, reliable data are not available. While men are the main perpetrators, this is one area in which women's rates of physical aggression may often be (at least in Western countries) as high as men's. (To be clear, this is evidence about parity in the *frequency* of physical aggression, not severity or motives, and does not concern other forms of abuse, coercion, and control.)

For example, as Helen Gavin and Theresa Porter report in their book *Female Aggression*, in a study of sixty-two hundred physical assaults between married partners living in Detroit, Michigan, the wives were more often the perpetrators of physical assault, injuring husbands through the use of weapons such as knives and guns. Another group of researchers reported on the frequency and characteristics of intimate-partner violence in six European cities: London, Budapest, Stuttgart, Athens, Porto, and Östersund (in Sweden). The researchers did not find that women assaulted their partners more often than men did, but they did find that "within each city, men and women presented equivalent prevalence of victimization and perpetration except for sexual coercion, more often perpetrated by men."

When I first learned of this evidence, I was skeptical. It ran contrary to everything that I thought I'd learned about domestic abuse, and it was hard to imagine women as significant perpetrators. But I hadn't looked closely enough, and when I did, although the evidence was uncomfortable, what I found made sense.

While women and men may not differ in their rates of physical aggression toward partners (or ex-partners), when women are aggressive, they are less likely to inflict serious physical damage. When a woman targets a partner, and throws plates, slaps, punches, or kicks them, the partner's injuries will be less serious, on average, than when a man carries out the aggressive act. This is particularly true in heterosexual relationships, in which there is a relatively consistent asymmetry in size and strength.

The reason that men are more likely to severely injure the targets of their aggression than women might not just be because they have bigger, stronger bodies—psychology might matter, too. Empathy is our ability to understand how others are feeling, and men are less able to do this than women, across cultures. This is a widely replicated and consistent finding, and it's not true just of human males and females. In chimpanzees, bonobos, gorillas, elephants, dogs, and wolves, researchers have observed that males engage in lower rates of behaviors related to empathy, like caregiving, cooperating, helping, and comforting. Reduced empathy might not only exacerbate the effects of greater male strength; it might also help to explain why men more often use lethal weapons, like guns, against partners. In any case, although rates of intimate-partner violence may be roughly equal between the sexes, the outcomes of these aggressive interactions are not. Men cause more damage and dominate in the most extreme form of intimate-partner violence, homicide. Worldwide, women are six times more likely to die at the hands of an intimate partner than are men.

The motivations for physical aggression toward a partner appear to differ. Both sexes can become violent when they fear a loss of a mate's fidelity, but men are more likely to use violence as a means to try to keep a mate from straying. And in all parts of the world, when women seriously injure or murder their partners, it is more likely to be in response to a history of threats and abuse to themselves, their children, or other family members. Women's motivations are more often self-defense.

Women, like all female animals, can physically be aggressive when they need to be. And they need to be aggressive when their lives, the lives of their children, or their future reproductive success are at risk.

MEAN GIRLS

Humans are strange animals in many ways, one of which is our creativity in causing each other pain. Not all human aggression involves a direct confrontation, like shouting "You're a pussy!," raising a fist in someone's face, or clashing antlers—all examples of what's called "direct aggression." "Indirect aggression," unique to our species, involves using language to get others to do your dirty work, like spreading gossip about a supposed friend or colleague and orchestrating their expulsion.

If you know anything about high school girls, this sort of aggression will be familiar. In junior high school, the alpha female of my friend group orchestrated the exclusion of one of my childhood friends from the group. I'm ashamed to say that I failed to confront our leader in an effort to stop it. I never understood how awful it was for my friend until we met up at a high school reunion, when she described how traumatic that event had been. (There was a silver lining—it helped her make new and more faithful friends.) Of course, boys and men indulge in this sort of viciousness, too, but girls and women seem to have an affinity for it.

Threats to one's children bring out a strong form of aggression in female animals. This morning on the way to school, I'll admit, I hollered at a driver who ran through a stop sign as my son and I attempted to cross the road. Maternal aggression helps to serve the reproductive goals of female animals—humans and nonhumans alike. But it does not appear to be linked to testosterone. In fact, studies on nonhuman animals show that the likelihood of maternal aggression is increased by the hormones of pregnancy and lactation. Because female aggression generally serves different purposes than male aggression, it tends to be modulated by different hormones.

If you define aggression broadly, so that it includes the indirect, maternal, and intimate forms discussed above, you can make a pretty good case that (human) females can be as aggressive as males. And there is clear evidence that women are just as disposed to anger as men. But if you define aggression more narrowly, as the kind of aggression that puts the perpetrator at physical risk, like inflicting bodily damage through acts such as headbutting, rape, and murder, there's no contest. Men win, hands down.

MURDER BY NUMBERS

We've already seen that "aggression" covers a multitude of diverse behaviors. Yelling in anger when someone steals your parking space, ostracizing a friend, threatening to punish a child, and plotting an assassination are all acts of aggression, but otherwise they seem to have little in common. How should aggressive acts be classified? I've already mentioned some ways of doing that, for instance, distinguishing between direct and indirect aggression, and researchers have proposed many more.

There's one more way to categorize aggression that's important to this discussion, and that's the distinction between "reactive" and "proactive" aggression. To see the difference, imagine you come home in the middle of your workday, walk into your bedroom, and discover your partner all twisted up in the sheets—hair tousled, naked, and with company. You feel your cheeks growing hot and your heart thumping in your chest. You scream expletive-filled threats and throw the framed photo of the two of you at their heads. That's reactive aggression.

In an alternative scenario, you act with cold calculation. Instead of throwing the photo, you whip out your cell phone and take one. Later you exact your revenge by posting it on Instagram and "doxing" (revealing private information about) your partner's lover. That's the second kind, proactive aggression. Neither sex has a monopoly on either kind.

As Richard Wrangham points out in his 2019 book *The Goodness Paradox: The Strange Relationship Between Virtue and Violence in Human Evolution*, reactive aggression is more likely to occur between two individuals (like the two headbutters at the start of the chapter), and proactive aggression is more likely to be perpetuated by groups of people or even institutions. He describes a type of "coalitionary" proactive aggression that is at the heart of such atrocities as war, torture, executions, slavery, and massacres.

Evidence linking testosterone and proactive aggression is sparse—although the role of the nervous system in proactive aggression is an active area of research. In this chapter, I'll be focusing on male competition for status, mates, or the resources to attain them, which tend to involve reactive aggression. As we'll see, for this kind of behavior, the evidence strongly supports a link to T.

MEASURING AGGRESSION

Measuring aggression is no simple matter. We get our best data on how animals behave when we observe them in the natural habitats to which they were evolutionarily adapted. But that's difficult when it comes to ourselves because we did not evolve in response to living in tight quarters, interacting with strangers on a daily basis, spending most of our time in school or at the office, being on Tinder, or getting our nourishment at McDonald's. What's more, eyewitness reports of aggression can be biased, especially when the witness was a participant. (One partial remedy for this problem is to use many different sources, such as reports from peers, or from parents and teachers in the case of children's behavior.) We can, of course, conduct experiments in the lab, where aggression can be measured more objectively, and also carefully manipulate hormone levels or the type of provocation. But it's sometimes not clear how to extrapolate from behavior in these artificial environments to behavior in the outside world.

To some extent, these problems can be overcome by studying

violent crime. While general crime statistics vary in their reliability, violent crimes are more likely to be accurately reported and recorded. The crimes aren't always solved, but the clearance rate for violent crimes is a lot better than for lesser offenses. This is especially true for murder, which is less susceptible to biased reporting. A nonlethal fistfight doesn't get misclassified as a murder, and murders almost always make themselves known.

Violent-crime statistics are also useful for evaluating the evolutionary basis of sex differences in aggression because they are available from diverse regions and cultures and across different time periods. These kinds of statistics provide only a limited picture, though, because they concern only the more extreme forms of physical aggression. It's rather like looking only at extreme forms of height. Out of one hundred people over seven feet, you'd be lucky to find a single woman. More than 99.9 percent of seven-plus footers are men—a massive difference! But if we move closer to sex differences in average height, the overlap is much greater. Out of one hundred people over five feet five inches, about twenty-two will be women. Relative to the differences at the extremes, sex differences in height closer to the average are far less imposing.

Sex differences in aggression (or almost anything else) work similarly. Most people live their whole lives without ever committing a murder or beating someone up, but many men and women have pushed, verbally threatened, or thrown something at another person. Those more typical aggressive acts don't usually make it into the crime statistics, and the differences between men and women here are much smaller than for violent crimes.

With those qualifications in mind, let's consider the painful reality of aggression at the extremes. Across distant times and places men murder and physically and sexually assault vastly more than women do. Men commit about 90 to 95 percent of all murders worldwide, and they most often kill other men. The women they kill are usually their wives, girlfriends, ex-wives, or ex-girlfriends. When men kill women their motivation is often sexual jealousy, and they kill as punishment for abandonment or perceived infidelity.

You may be wondering, quite reasonably, how it could ever be

adaptive for a man to kill his wife or girlfriend, since doing so elimi-
nates what evolutionary theory predicts he should be most motivated
to protect. Men who physically intimidate and coerce an otherwise
reluctant mate to stay sexually and emotionally loyal may resort to
killing as a kind of overextension of efforts to prove the credibility of
the threat. As the two preeminent scholars of the evolutionary bases
of familial homicide, Martin Daly and Margo Wilson, explain it:

> Men strive to control women by various means and with variable suc-
> cess, while women strive to resist coercion and maintain their choices.
> There is brinkmanship in any such contest, and homicides by spouses
> of either sex may be considered the slips in this dangerous game.

Statistics for other kinds of violent crime, like assault, are not as
accurate as those for homicide. Rape, in particular, may be seriously
underreported and underprosecuted, and its legal definitions vary

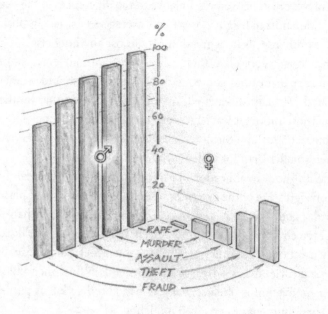

Sex differences in worldwide crime rates

widely. (No matter the definition, women are almost never arrested for rape.) But it's clear that whatever the precise figures for violent crime, men dominate in all categories. Taking arrest rates as a rough proxy for the composition of offenders, men commit 80 to 85 percent of violent crimes in the United States. Worldwide figures are similar. Although the ratios for each country vary, men commit about 90 percent of physical assaults (injuring another person on purpose) but only about 80 percent of thefts (when the theft involves more physical risk, like stealing a car or breaking into a home, the number is higher). And when crime involves no physical risk, like forging checks or embezzlement, women are responsible for a larger portion than for the violent crimes. Worldwide, men commit more fraud than women, comprising (with wide cross-cultural variation) roughly 70 percent of the perpetrators.

The more risky, extreme, and cruel the violence, the larger the sex difference, and the greater the proportion of male offenders. As one of the world's leading aggression researchers, the psychologist John Archer, puts it: "There is a difference in the degree to which men and women escalate aggressive exchanges to dangerous levels." Men may end up butting heads or killing, while women are more likely to put the brakes on at dirty looks and yelling, or pushing, kicking, or slapping.

High rates of male aggression appear to have long been a feature of the human condition, tracing back to our distant ancestors. Ancient fossil skulls show evidence of violent conflict, including cracks and holes that are thought to result from attacks with clubs, rocks, or spears. This type of damage is far more common in male than female skulls. And among the hunter-gatherer populations remaining today, the sex differences in homicide rates are comparable to those in the rest of the world: men commit almost all murders, mostly against other men.

AGGRESSION AND SEXUAL SELECTION

In most species, males do not all have the same level of sexual access to mates. It's just a fact of life that there are reproductive—and thus evolutionary—winners and losers. Wisdom 11, the red deer stag from the previous chapter, was a big winner. And the more he wins, the more others lose: there are only so many fertile hinds to go around. Hinds, however, play no such zero-sum game. While they may compete for resources like food that impact reproductive success, they don't need to fight among themselves for mates. That would be a risky strategy with little benefit. In other animals, like humans, females do compete for mates, but the competition tends to be less physically risky than for males.

Often (although not always), physical aggression has an evolutionary dividend for males that it does not have for females. If males can use aggression to win mating opportunities and exclude other males from accessing them, or to prevent females from straying, then the force of sexual selection will favor genes that promote this behavior.

Is there evidence that men are adapted for potentially violent mate competition with other men? Yes—quite a lot of it. Men are bigger and stronger than women, are maximally aggressive at prime reproductive age, and take greater physical risks (for example, vastly more young men die in car accidents and from drowning than women). Boys, like many other young male primates, engage more than girls in rough-and-tumble play, useful practice for real physical contests later. Men are overrepresented in violent contact sports like mixed martial arts and boxing—and this is true from Africa to Asia to the Americas. Players of violent video games also tend to be male, and men fantasize about fighting others much more frequently than women do.

Being prepared to use violence has costs—perhaps even your own life. Evolution will only favor an aggressive disposition if it brings reproductive benefits that outweigh the costs of injury. If men are adapted to compete with other men, we would expect that some would be reproductive winners (like Wisdom 11) and others would be losers. Women's reproductive outcomes, in contrast, should cluster more tightly together. And in fact, across a diverse range of human societies (with a few exceptions), this is what we see. There is more variance among men than women in the number of offspring they leave behind. In modern Western societies this sex difference in reproductive variance can be relatively small, but in more traditional societies like the Aché hunter-gatherers of Paraguay or the Kipsigis agro-pastoralists of Kenya, it can be great. And in polygynous societies, in which a small number of men can have two or more wives, the distance between the evolutionary winners and losers will be even greater.

Sexual selection adapts animals for mating competition, and testosterone helps male animals to take advantage of those adaptations by coordinating male reproductive physiology and behavior. There is no reason to think that these influences have been mysteriously switched off in the human lineage. Even without looking at the details of how T modulates aggression in men, there is a strong initial case that it is a significant factor in explaining sex differences in violent behavior.

MEN, MALE CHIMPS, AND SPARROW DADS

Unlike red deer, spiny lizards, and song sparrows, our closest ape relatives, chimpanzees, are not seasonal breeders. (Neither are the more playful and peaceful bonobos which, together with chimps, are our closest living relatives.) The testes of male deer, lizards, and sparrows are shut down for long periods when there are no fertile females around, enabling the males to live relatively peaceful and sexless lives for much of the year. But the testes of men and adult male chimps are always on the go. These guys need to be ready to mate and to compete but must also avoid the costs of maintaining T levels that are cranked up to the max. So the testes stay on alert. They make enough T to produce lots of sperm and maintain secondary sex characteristics like extra muscle and a propensity to fight, and stay prepared to ramp up T production to fuel intense competition should the situation call for it. Instead of using seasonal cues to move testosterone around, in humans, chimps, and other nonseasonal breeders, T levels are sensitive to threats from other males. And in chimpanzees, males threaten and fight much more in the presence of females who are close to ovulation and able to conceive.

Female chimps advertise their fertile condition with a "sexual swelling"—a large, fleshy protrusion on the backside. A female with

Estrous female chimp, male looking on

this gigantic signal is sexually attractive to male chimps. Research from the Kanyawara field site in Uganda (where I spent my time with the chimps before coming to graduate school), conducted by Martin Muller and Richard Wrangham, shows that the T levels of adult male chimps rise to their highest levels when females are in estrus.

The same pattern holds in the red deer, song sparrows, and spiny lizards: T, aggression, and sexual behavior all move in concert with female fertility and the need to compete for mating rights or the resources required to attain them.

But in women, ovulation is concealed. That is, women don't advertise their fertility to the males of their species with a temporarily tumescent backside, irresistibly sexy smell, or obvious sudden increase in flirtatiousness. (Some evidence does suggest, however, that men can subconsciously detect subtle changes in women's smell or patterns of movement around ovulation.) Instead, women are at peak sexual attractiveness throughout their reproductive years, without giving away exactly when they are able to get pregnant during that time. (This helps to explain why men stick around even when women can't get pregnant, unlike most other mammals, but that's another story.)

Concealed ovulation is one important difference between ourselves and chimps—and 95 percent of mammals. And there's another: in humans, children are more likely to survive and thrive when their fathers provide care, which helps to explain why many men invest time and energy in their children. And, just like song sparrows, a human dad's T levels tend to drop when they are in a romantic pair bond and involved in caring for their offspring. The links between T changes and behavior in men aren't as tight as in nonhuman animals, but the evidence is consistent with the idea that this decline in T helps direct attention away from rivals and new sexual prospects and toward one's mate and children.

STATUS COMPETITION

Evolutionarily speaking, violence should be reserved for when the payoffs are worth it—a tactic of last resort. Both the weak and the strong benefit from a system that lets each animal know, in advance, what their chances are likely to be against another animal, and who should back down, confront, or challenge whom: a system that would minimize actual fighting, in which each animal knows their place and can read and respond to the signals that tell them when to be deferential and when to be bold. One that allows a group of animals to enjoy the benefits of living together—raising families, finding food, goofing around, defending territory, having sex, and cooperating as necessary. In other words, a status hierarchy.

Recall Daemon Fairless's encounter with the drunk man on the train. It was typical of much violence between men: a relatively trivial dispute escalates, with neither side backing down, and finally the gloves come off. It could have been worse—many murders of men by men start in this way. It could have been better—countless disputes like this quickly fizzle out.

Two men stupidly fighting over nothing? In one sense, yes: they weren't fighting over anything that someone could observe, like a bag of gold or territorial boundaries. But in another sense, no. They were fighting over something that is both intangible and immensely important: social status. The more respect and deference we pay to an individual, the higher his or her social status. The drunk man flaunted the rules that high-status middle-class individuals like Fairless abide by—no yelling and swearing, no scaring the other passengers. And no coming anywhere near my wife while doing so.

Many primates have similar status totem poles. To acquire or maintain a high-status position, males rely on not just social skills and the ability to build coalitions but also the credible threat of violence. And in situations in which males use aggression to dominate others, a status hierarchy serves to reduce the intensity and frequency of fights. Knowing your place and signaling your acceptance of it to dominant individuals—chimpanzees have a "pant grunt" for this purpose—means that community members don't have to duke it out every time a conflict arises. Everybody knows who is most likely to win any fight over that estrous female or prime feeding spot. But failing to obey the rules demands a response from those higher up, whose interests lie in maintaining the status quo. After all, those at the top are winning, and not just the best food, sleeping spots, money, or power; they are winning mates and beating their competitors at getting their genes into future generations.

So why did Fairless engage violently with the drunk man on the train? Why take such a risk with someone who was a stranger to him, whom he would never see again? Fairless's position on the social ladder was not really threatened. It's not that the drunk man might have realistically stolen Fairless's job, let alone his wife. Why didn't Fairless just let it go?

A (speculative) answer lies in our distant past. Human beings evolved in response to a social environment that bears little resemblance to that of today. Estimates (based largely on contemporary foragers) suggest that hunter-gatherer societies averaged about one thousand people, at least half of whom were children. People within a given society resided in smaller, more fluid camps, sharing land, language, and customs. On average, these camps consisted of about fifty people. Only rarely did they number more than three hundred.

Strangers are therefore a novel feature of the modern human environment. Back in the day, everyone in a given society knew everyone else. Men understood their social status not only within the context of their residential camps, but also within broader society, and the hundreds of men with whom they would have interacted frequently across their lifetimes. This lack of anonymity means that any particular behavior could have far and long-lasting reputational consequences: backing down from an evenly matched competitor could damage hard-won status.

Even though that train was full of strangers, Fairless and the other man responded the way that evolution and testosterone shaped them to respond, as though they were back in the ancestral group, with status to be determined and without the option to make a clean exit at the next stop.

RAPID T CHANGES

Relative to females, the human male is exposed to high levels of T in the womb, in early infancy, in puberty, and throughout most of adulthood. Testosterone in fetal and pubertal development sets up a male's brain and body to allow him to respond to high levels of T in adulthood. T gives men a tool to direct energy in ways that help to overcome their specific limitations to making babies—that is, finding and retaining mates.

T levels in men normally change throughout the day, declining from 40 to 50 percent or more from morning to night, and they also rhythmically rise and fall with each pulse of T from the testicles, which is sent out into general circulation about every sixty to ninety minutes. But in many male animals, including men, T levels can change more rapidly—within a few minutes—in response to social interactions that are relevant to reproduction. The point of these socially induced fluctuations in T appears to be to help the male react adaptively (that is, in ways that would have increased his reproductive success in our evolutionary past) to present and future social situations that are relevant to mating and status competition. Should he react with boldness and dominance, or fear and submission? Fight or flee? In the heat of the moment, he can't pause to carefully weigh the pros and cons of engaging in a battle with a particular challenger. This is where T may come to the rescue, changing in ways that help men to rapidly assess their chances of competitive success, increasing the likelihood of making an adaptive response.

These short-term changes in T have been observed in men in a number of experimental and natural settings, one of which exploits that very common way modern men have of competing with one another—professional sports. A soccer or boxing match might not seem at first glance to bear on one's mating prospects, but sports provide an opportunity to test and improve one's competitive capacity, build coalitions, and ultimately, to increase social status.

WINNERS AND LOSERS

On July 17, 1994, at the Rose Bowl in Pasadena, California, Italy and Brazil squared off in the final game of soccer's World Cup, the world's biggest sporting event. This gave some behavioral endocrinologists from Georgia State University an idea.

Prior to the game, equipped with test tubes and participation forms, they went out on the town, found a crew of Italian male fans in a pizzeria and a crew of Brazilian male fans in a bar, and in both cases got some of them to produce saliva samples before and after the game. It was all part of a simple experiment to determine T levels.

After ninety minutes, neither side had scored. This was soccer, after all. Extra time brought no goals, either, so the game went to a heart-in-the-mouth penalty shootout. The world held its breath, and players from each side took shots on goal, one by one.

And who won? The researchers could have figured it out just by consulting the testosterone levels in the men's spit samples. Levels of the Brazilian fans stayed about the same or increased at the end of the game, and they dropped for the Italians. Brazil wins! Many studies have found similar results, in which T rises prior to a competition in both the eventual winners and losers, but T remains elevated for longer in the winners than the losers. These kinds of T responses have been demonstrated even in men who compete in unathletic events like chess and video games. While this effect occurs regularly in humans and nonhuman animals, it's variable. Many factors affect how T responds to competition, including the circumstances of the experiment (like whether the competition is orchestrated in the lab or in the "real world"), how much someone cares about winning, and levels of other hormones.

Winning or losing a physical contest provides evidence of an animal's fighting ability. A winner should capitalize on his abilities, and a loser should be extra cautious in responding to a challenge—getting beaten to a pulp won't improve your dating prospects. This is the so-called winner-loser effect, which has been demonstrated in insects, fish, birds, and mammals. And at least in nonhuman species, the effect is clearly mediated by testosterone (or chemical variants of T that are found in some nonhuman animals).

For a well-studied nonhuman example, take male Syrian hamsters. My son has one for a pet—Ringo, named for the fourth Beatle. If we put a bigger male into Ringo's cage, the two would duke it out, and Ringo, who is timid, would likely lose and submit. The next day poor Ringo would still be reeling from his previous defeat, and if we put even a scrawny little unaggressive hamster in his cage, Ringo would simply cower in fear and do his best to defend himself if attacked. This effect lasts about a month in Syrian hamsters. And we know it involves a drop in T.

Studies of male Syrian hamsters in these situations have shown that if the loser is given extra T shortly after a defeat, it will block the typical loser response, and he'll feel like a winner and continue to aggressively defend his home territory. A dose of T after a loss reduces feelings of stress and fear and increases the motivation to fight rather than flee when confronting future challenges.

Of course, overconfidence is generally a bad idea in the wild. Animals with a history of losing need to be extra cautious to stay alive, and a drop in testosterone helps them do that. T changes provide information to the animal's brain by changing its sensitivity to other hormones and neurotransmitters (chemicals that allow neurons to communicate) that are involved in the perception of and response to threat. Next time the losing animal is threatened, he might feel a

heightened sense of fear, pain, or anxiety, which reduces the likelihood of an aggressive response, but increases likelihood of survival. If
Ringo had won the first fight, he'd have experienced a T rise, and he'd
be more likely to feel bold in the face of the next threat, and motivated
to defend his territory against future intruders. And in nature, being
a dominant male and using aggression judiciously lead, on average, to
increased reproductive success.

PERSONALITY MATTERS

Whether playing hockey, playing a video game, arguing, or even
throwing punches, men may feel that their status or reputation is on
the line, driving them to do what it takes to be victorious. There's no
doubt that testosterone levels often respond, in the short term at least,
to competition (which may sometimes be aggressive) between men.
This fits with the Challenge Hypothesis, which predicts that male T
levels rise in response to threats to status or resources. But much is still
unclear: the purpose of such T changes, the biochemical mechanisms
involved, the characteristics of the men in whom the effects occur, and
the kinds of social interactions that can influence T levels.

Fortunately scientists are making some fascinating discoveries
about how this might all work. A team of reseachers, led by Canadian
psychologists Shawn Geniole and Justin Carré, has conducted several
studies that explore the effects of experimentally elevated T levels on
aggression in men. "Dominance-oriented" men, who are "assertive,
forceful and self-assured," are motivated to achieve status and power.
And if that trait is tied to another—lack of self-control—then a provocation is more likely to result in an aggressive response. The researchers
found that dominance-oriented men who are also impulsive—lacking
self-control—are the ones who tend to respond to an elevation in T
(as might occur when a man anticipates competition) with aggression.

So T may increase aggression, but only in a subset of men. This
finding helps explain the lack of consistency in previous studies. If
the study design includes only a small number of men (which is usually the case, since these studies are complex and expensive), then

a significant result is less likely, because there might not be enough experimental subjects with the right personality traits.

To more fully explore the relationship between T and aggression, these researchers ran a follow-up experiment, the largest of its kind to date. They recruited over three hundred men, who, among other tasks, filled out a questionnaire about their personality and played a computer game. Not Grand Theft Auto, but a game that is a common and well-validated tool to measure aggression in the lab, the academically titled Point Subtraction Aggression Paradigm, or PSAP.

The men were led to believe they would be competing against another player (a man) whom they had seen in a video. (In fact, they played against a computer program.) The goal of the PSAP is to press certain keys to earn as many points as possible, which could be exchanged for cash at the end. Instead of earning points for themselves, though, participants may choose to press a different key that reduces the other "player's" points, which doesn't benefit the participant and only harms the other "player." Populations known to be aggressive, like violent criminals, in addition to other people who describe themselves as aggressive, tend to earn fewer points and less money in the PSAP since they spend more time stealing points from their "competitors." In essence, these men pay just to behave aggressively in response to a provocation.

Before the game began, each participant was given a gel to squirt into his nostrils. It contained either testosterone, which raised blood T-levels within fifteen minutes, or no active ingredients (a placebo). Over the course of the experiment, participants did no better than chance at guessing whether they had received testosterone or the placebo.

The results confirmed previous findings. The men who received the T gel behaved more aggressively—that is, they stole more points from the other guy. And like the previous experiment, the effect occurred primarily in the dominant, impulsive men.

GENES MATTER

But that was not all. The researchers didn't just look at personality—they also looked at genes, specifically, the androgen receptor gene.

Remember my student Jenny from chapter 3? Her androgen receptor gene contained a mutation that put her androgen receptor out of business. In partial androgen sensitivity syndrome, the receptor works, but it is not as responsive to androgens as a fully functioning receptor. So it's not all about T level: the effectiveness of the androgen receptor matters, too. The same point holds for ability of the T-bound androgen receptor to control the rate of protein production from its "target genes," like those that promote beards, muscle, or aggression. As it turns out, even among the typical, fully functional androgen receptors, some are really efficient at controlling transcription from target genes and cause more proteins to be produced, and some are less so.

The researchers wanted to see if a given amount of T would preferentially increase aggression in men with relatively efficient androgen receptors, so they collected (used) mouthwash from the men in the experiment, from which they harvested their DNA.

The span of DNA that comprises each androgen receptor gene has something in it called a "CAG repeat," which is a segment of C+A+G "letters" that repeats—from eight to thirty-seven times. Fewer repeats means the receptor is more efficient, and more repeats means it's less so. All else being equal, the people who should have the greatest response to a given amount of T are those who have the lowest number of CAG repeats in their androgen receptor gene.

The length of one's CAG repeat turns out to be associated with all sorts of things, such as the probability of getting prostate cancer (more likely with fewer repeats), pregnancy outcomes (spontaneous abortions are more likely with fewer repeats), and even one's ethnic background.

And now we know one more thing that the CAG repeat predicts—aggression in response to testosterone. Among the men who were dominance oriented and impulsive, those who had a short CAG repeat, and thus a higher sensitivity to T, responded with more aggression (they stole more points) to the T gel.

Furthermore, the men with shorter CAG repeats reported that they got more pleasure from being aggressive! This finding provides some insight into how T might encourage aggression in men. Elevations in T provide incentives, because T increases sensitivity to reward. For example, when mice are given a choice of which side of a cage to hang out in, they will choose the side in which they previously received a nice dose of testosterone. The parts of the brain that are rich in the neurotransmitter dopamine, and which influence motivation, are dense with androgen receptors. We tend to get a rush of dopamine when we do something adaptive (or that was adaptive in our evolutionary past), like eating something sweet, having sex, or intimidating a competitor. The dopamine helps reinforce that behavior: since it feels good, we are motivated to do it again. And animal studies show that T actually increases the amount of dopamine that is released in response to winning, in addition to increasing the number of its own receptors in the reward centers of the brain! All of these changes appear to make it more likely that winning animals will be more likely to confront future threats.

To sum up the lessons from this study: increasing levels of T appear to be motivating and rewarding in some men in the right circumstances, with the right personality, and the right kind of androgen receptor gene. That all these factors matter is not evidence that the relationship between T and aggression is weak; rather, it shows us that it's complicated, as is the research that looks into how the relationship works.

T has other effects that make severe aggression more likely. First, T appears to reduce empathy. If T levels are elevated, then motivation and reward are increased, and fear and the perception of pain are decreased, allowing animals to escalate fighting. Reducing T reverses these effects, suggesting that those whose T is lower or falling are nudged by pain and fear to do what's adaptive for them, and run or bail out. Similar patterns hold for nonhuman animals.

No matter the animal, testosterone's relationship with aggression is clearly not one of simple cause and effect but instead is modulated by factors like previous experience, personality, and one's position in a status hierarchy. In his book *The Trouble with Testosterone*, the Stanford biology professor and expert on the endocrinology of aggression Robert Sapolsky illustrates this point with a description of an experiment in a captive group of Talapoin monkeys. The monkeys are introduced to each other and given time to form status hierarchies. When experimenters increase one of the monkeys' T levels—"enough to grow antlers and a beard on every neuron in his brain"—he does do more chasing, grabbing, and biting. But what's interesting is who is on the receiving end of his increased aggression: the T'd-up monkey is not indiscriminately harassing anyone who happens to irritate him. Instead, he beats up only on those beneath him on the totem pole and remains polite toward the higher-ups.

T is not a potion that turns the meek into warriors or that causes rampant bellicosity. Its effects depend heavily on individual and environmental factors, and in humans especially, winning and achieving high status can often be accomplished without any physical aggression at all. T tends to do what the situation requires. As Sapolsky joked during a lecture, if you shot up a bunch of Buddhist monks with testosterone, it would lead not to violence but to random acts of kindness.

RAPID T MECHANISMS

We don't know how T is produced so rapidly when men find themselves in competitive situations. The rhythmic signal to produce T from the testicles originates in the brain, starting with the hypothalamus, which sends a pulse of gonadotropin-releasing hormone to the pituitary gland, directly below it. This signal stimulates the pituitary to release luteinizing hormone (LH) into the blood. That LH, in turn, needs about one hour to move from the brain to the testicles, where it tells them to produce and release T into the blood. Given that relatively slow, long-distance system, how T levels can rise within a few minutes of a social interaction remains, at present, a mystery. Research suggests one possibility: the hormones adrenaline and noradrenaline, which are released in times of psychological and physical stress (for instance, prior to competition or in the face of threat), could bypass the LH system and stimulate the release of T from the gonads, either directly or by increasing blood flow to them.

There's another puzzle as well, and that is how T could act quickly when it reaches cells, particularly neurons. As I explained in chapter 3, T normally activates androgen receptors inside cells and ultimately affects the expression of certain genes in the cell nucleus. That entire process takes time and, as far as we know, does not affect behavior in a few minutes. This means that the finding that the length of the CAG repeat is associated with an aggressive response to testosterone isn't as straightforward as I made it out to be. The mechanism can't simply involve more efficient transcription of target genes, because that would take too much time. Some new and exciting research, however, suggests that T could have important, and more rapid, "nongenomic" effects—meaning, actions not inside the cell on gene transcription, but instead right on the cell surface, more like a neurotransmitter or a protein hormone. We are getting into the mechanistic weeds here. But speaking as someone who was turned off by science until college, asking seemingly simple questions about T's action helped me to realize how interesting even the nitty-gritty neurological, hormonal, and genetic influences on behavior can be. Further research in this area will go a long way toward helping us understand the ways in which T might modulate aggression.

T AND WOMEN'S AGGRESSION

Women also compete for status, resources, and mates, sometimes fiercely. And such behavior is sometimes related to T levels in some nonhuman female animals. So shouldn't women also experience the highs and lows of T in response to competitive interactions? Certainly lots of researchers think so, and they design experiments to look for a T-mediated winner-loser effect that we sometimes see in men.

To complicate the picture, androgens are produced by the adrenal glands, which amp up activity in times of stress in both sexes. And competition increases stress. This introduces a new wrinkle in the interpretation of testosterone changes in the winner-loser effect, especially in women. This is because (in addition to problems in measuring female T), the adrenal gland produces roughly half of circulating testosterone in women. If we find that women's T rises in anticipation of competition, this might simply be a by-product of the adrenals responding to stress.

The scientific literature contains little evidence to suggest that T mediates female competitiveness. Of the studies on the winner-loser effect that include women, almost all fail to report changes in T that are related to status, changes in status, or winning or losing; that is, a role for T in competition that is similar to that in men.

This shouldn't be surprising. From an evolutionary point of view, males and females are expected to respond differently to competitive threats, as we've already seen. Facing them with physical aggression can more often pay off for men, and women have less to gain and more to lose reproductively from engaging in violent competition. They should be less motivated to engage in physical confrontations in the first place, and they should be biased toward a quicker withdrawal from aggressive interactions as they heat up. Given the adaptive and physiological differences between the sexes, it would be astounding if men's and women's testosterone production responded in the same way to competition.

None of this means that women don't care about winning, even as much as men in some cases. There are plenty of examples of intensely competitive women in business, sports, or school. There are other hormones involved in competition, like cortisol (which is now getting much-deserved attention), and in women it's quite possible that estrogen and progesterone also play a mediating role. Let's get away from the idea that in order to validate stereotypically masculine behavior in women, we must show that T works the same way in both sexes. Hopefully we'll see more research on the hormonal mediators of women's competition in the future.

ENVIRONMENT MATTERS

The fact that Daemon Fairless and the drunk man lived in a society with a proprietary set of customs and social standards surely had something to do with how events unfolded on the Toronto subway that New Year's Eve. Perhaps unsurprisingly, Daemon's headbutt was not met with disapproval from his fellow passengers, or even from the Toronto cops. "The police gave me a subtle nod of approval for defending myself, even as they pointed out the obvious risks . . . there was an understanding between us—I'd taken the risk to protect my wife."

Many societies positively prize actions like Fairless's—men are expected to use physical aggression to defend their families and reputations. This kind of "culture of honor" is found in the American South, which has historically higher rates of violent crime than the North. Describing how this affected boys, the historian David Fischer wrote:

> From an early age, small boys were taught to think much of their own honor and to be active in its defense. Honor in this society meant a pride of manhood in masculine courage, physical strength, and warrior virtue. Male children were trained to defend their honor without a moment's hesitation—lashing out against their challengers with savage violence.

If Daemon and his drunk adversary had been riding a train in, say, Singapore, then I'd bet the ranch that any altercation wouldn't have ended in headbutting. (And anyway, Singapore's young men don't tend to get drunk—in public, anyway. Public drunkenness is a serious offense, a fact that many visitors have discovered to their cost.) The rate of violent crime in Singapore is minuscule compared to Jamaica, the United States, or even relatively peaceful Canada, whose assault rate is about fifty times greater than Singapore's. Singapore has the lowest murder rate in the world, alongside Japan.

Why is Singapore so different? The government doesn't pump pacifying chemicals into the water supply. The explanation presumably lies in the Singaporean culture, which is one of law-abidingness, strict discipline in families, and lack of poverty, plus harsh criminal penalties, among other factors.

Rates of violent crime don't vary just from country to country, they also vary over time. As Steven Pinker has documented in *The*

Better Angels of Our Nature, the homicide rate in Europe fell astonishingly steeply from the thirteenth century on, from as much as 100 per 100,000 people per year, to the present rate of about 1 per 100,000. The explanation for the reduction in violence is not in changes in our genes but in centuries of large cultural and social changes, including the monopolization of violence by the state.

Despite these variations, one thing that remains constant is the gendered pattern: across vast reaches of time and space, men are more violent than women. That is a striking fact that requires an explanation, and the most parsimonious one, consistent with vast amounts of evidence from across the animal kingdom, invokes sexual selection and its handmaiden in males—testosterone.

Although humans are not held captive by evolutionary forces, genes, or hormones, we still experience their profound influences. But we are unique among animals in some important respects, including our ability to carefully consider the consequences of our actions and inhibit our baser instincts. The more we understand the forces that shape us, the more control we have over how we behave.

As I have stressed throughout this book, behaviors that show sex differences are often heavily influenced by culture—and aggression is a clear example. Laws and cultural and social norms can push physical aggression up or down. We can hope that social changes will reduce violence—perpetrated largely by men—yet further. But one can't solve a problem if one misunderstands its causes. Frank talk about T will help us appreciate how changes in the environment can rein in problematic male behavior. It is within our power to close or widen sex differences in aggression—but the underlying tendencies producing those differences precede culture, and they exist because of testosterone. No good can come from denying that.

8

GETTING IT ON

OF RATS AND MEN

Let me tell you a story that every student of endocrinology hears sooner or later. Alas, it probably isn't even true—but, as I'll explain, it's retold for a good reason.

In the 1920s, U.S. president Calvin Coolidge and his wife were taken on separate tours of an experimental government farm. When Mrs. Coolidge visited the poultry area, she noted that the cockerel was mating frequently, and she asked the poultry hand how many times a day this occurred. His answer was "dozens of times." "Tell Mr. Coolidge," she replied. When his wife's message was conveyed to him, the president asked, "Was it the same hen each time?" "No," said the poultry hand, "it was a different hen each time." To which the president replied: "Tell Mrs. Coolidge."

My students do laugh at this story, but that might just be in response to my pathetic attempt to tell a joke (or they are angling for a higher grade). Funny or not, it illustrates a genuine phenomenon: renewed sexual interest and capacity to respond to a new potential sexual partner. This phenomenon is called the Coolidge effect. Researchers have demonstrated it in a number of animals including rats, fish, sheep, cattle, monkeys, and chimpanzees. In one of my favorite such experiments, researchers place an adult male rat in a cage, which is divided in two by an opaque barrier. The hair on his head has been shaved, and sticking out of the top is something called a "microdialysis probe." It's not a good look. The large, strange lump sprouts long wires that allow the rat to move around freely while researchers monitor levels of the neurotransmitter dopamine in his brain. Dopamine has many functions, and an important one is to increase motivation and reward; elevated levels suggest that an animal is anticipating a reward and is motivated to pursue a certain goal.

At first, the male is just hanging around with a funny-looking thing sticking out of his bald head. His dopamine levels are normal, at baseline. But then everything changes. An estrous female rat is placed on the other side of the barrier dividing the cage. They are both on alert, sniffing around like crazy, getting the information they need from the other's scent. Health status, what they had for lunch, hormone levels, and more—she can tell whether he's a sexually mature male with high T, and he can tell the phase of her cycle (whether she's able to conceive). Bang, up jump his dopamine levels, by about 50 percent.

The evil scientists keep the poor rats separated. It's like being really hungry and then having someone put a plate of your favorite food in front of you, the smells wafting up your nose, with your hands tied behind your back. This stage of animal sexual behavior is appropriately called the "appetitive" phase, and the part where animals actually have sex is called the "consummatory" phase. The neural systems (including dopamine changes) that motivate the pursuit of sex and the reward of getting it on are very similar to those that motivate the search for food and the satisfaction that comes with consuming it.

When the barrier is finally removed the rats get down to business. The male rat's dopamine now doubles from baseline and stays very high until he ejaculates, when it falls somewhat. But it's not over! He's still into her, and they continue to mate, with the male ejaculating a handful of times, taking breaks in between. And with each ejaculation, his dopamine levels go down a bit. When he's had enough, his dopamine levels return to baseline.

The female rat, I should say, is not a passive sexual partner. When she can control the pace of mating, she has a higher likelihood of pregnancy; also, her dopamine rises more than it would when she is not in control, suggesting that sexual control is adaptive and rewarding for female rats.

After fifteen minutes of low dopamine in the male and no sexual activity, the female is removed. The male hangs around by himself for another fifteen minutes, after which the same female is introduced again behind the barrier.

Nothing happens.

His dopamine and sexual interest don't budge, even after the barrier is removed and the postcoital rats have free access to each other. He's sexually exhausted, his tank is empty. She is removed, this time for good.

But wait, there is something left! After another few minutes, researchers put a different female on the other side of the barrier. He gets a whiff of her fresh, estrous scent, and his dopamine starts to climb back up. When the barrier is lifted, he is again up and at 'em, albeit with slightly less vigor than before. And his dopamine shows it. It doesn't rise as much as the last time, but enough to provide the motivation he needs to pursue her, mate, and ejaculate again, which he does several times.

Sex may come naturally to us, but getting it, and getting it right, is a complicated feat. Think about it from the point of view of a male rat. He has a lot he needs to accomplish in order to successfully mate. For starters, he needs to decide whom to pursue. He's got to perform the actual movements of pursuit, and his advances must be accepted. His penis must become erect, he must mount the right part of her body, insert, thrust, and release sperm into her vaginal tract. Phew!

Dopamine, in concert with testosterone in the body and sexy cues in the environment, works to help the rat bring this off. Dopamine is crucial for the expression of all kinds of motivated behavior, like searching for food and water, or avoiding predators and moving away from things that hurt. It plays a central role in controlling movement, and coordinates the motivation to act with the movements necessary to do so. (In people with Parkinson's disease, low dopamine makes regulating movement difficult.) Dopamine plus T enables male rats to focus their attention on estrous females, rather than wasting time on mice or inanimate objects.

Exhibiting the Coolidge effect seems to be an adaptive trait in males of many species, meaning that the trait increases reproductive success. If more sex with the same female won't increase the chances of pregnancy, the male loses nothing by stopping. But if another fertile female comes around, that's another opportunity to leave offspring that may be too good to pass up.

Why am I starting this chapter, primarily about T's role in the differences between men's and women's sexual behavior, with the Coolidge effect in the rat? Well, insofar as the story about Mr. and Mrs. Coolidge is at all funny, it's because it plays on the stereotype of men as being more eager than women for new sexual partners. As I will explain, not only is that stereotype accurate, but T has lots to do with it.

Overall, the sexes do have lots in common in terms of their sexual preferences and mating behavior: both men and women feel strong sexual desire and seek partners who are good-looking, kind, healthy, intelligent, and honest. But when it comes to having sex, men want to do it more often, and with a greater number of people (we'll get to that shortly). Yet there is one sex difference in sexuality that is truly massive, although easy to overlook. It's the target of sexual attraction: the vast majority of men are sexually attracted to women, and vice versa for women. Testosterone helps the male rat home in on estrous females—what is its role in human sexual orientation?

I'll tackle sexual novelty and sex drive first, and sexual orientation later. Let's begin at the point where humans start to get sexy: puberty.

CRUSHES, KISSES, AND GOING ALL THE WAY

Prior to the surge of sex hormones in puberty, most kids think anything to do with sex is gross—possibly because they feel its pull and they're not emotionally ready to handle it. Beginning at ages six to seven in girls and seven to eight in boys, androgens are already on the rise. This happens well before the testes and ovaries get up and running, churning out their main products: sperm and eggs, respectively, and sex hormones. These early androgens are produced not by the gonads but by the adrenal glands (which, to remind you, sit atop the kidneys and produce cortisol, among other hormones). These "adrenal androgens" rise until our early twenties, when they begin their slow decline. The main adrenal androgen is a bit of a mouthful—dehydroepiandrosterone, more compactly known as DHEA. In both boys and girls, DHEA is released into the blood and is converted into low levels of testosterone (this conversion happens in "peripheral tissues" like the liver, kidneys, and brain). It is this adrenally derived T that causes the first appearance of slight pubic hair, acne, and body odor in both sexes. It may also be responsible for the earliest of crushes and hints at sexual feelings. But these adrenal androgens aren't enough to stimulate the growth spurt, and true puberty doesn't start until a few years later.

A light refresher about puberty. It begins with the brain and the gonads communicating through the hypothalamus-pituitary-gonadal (HPG) axis. In the brain, the hypothalamus sends regular signals, in the form of gonadotropin-releasing hormone (GnRH) to the pituitary, which responds by sending luteinizing hormone (LH) and follicle-stimulating hormone (FSH) through the blood to the gonads. LH and FSH stimulate the ovaries and the testicles to produce sex hormones and eggs or sperm. (See the figure on page 117 in chapter 5.) These changes in bodies and feelings aren't just happening in Western cultures, but in girls and boys all over the world.

Those sex hormones—estrogen and progesterone in girls, and testosterone in boys—kick off puberty, indicated by the increased growth of pubic hair, the formation of breast buds for girls, and the enlargement of the testicles for boys. Girls enter puberty about a year before boys (about ten and a half versus eleven and a half, with lots of variation depending on culture, environment, and ethnicity) and will often get their first period before boys their own age have even started puberty. The boys' delayed start is common in males of other species in which the adult males compete for mates. A longer period of juvenile growth gives young males time to grow to a larger body size before T begins to redirect its energy toward developing secondary sex characteristics, like enhanced muscle mass, that will support mate competition.

As the sex hormones increase, sexual and romantic interest intensifies. Most early adolescent boys and girls report having a crush on someone (girls have more crushes). Those crushes soon start to take on a more sexual tone. Thankfully, most kids that age aren't sexual with other people, but they are with themselves. By the time they turn fourteen, around 90 percent of boys and 20 percent of girls (in the United States) have masturbated. The first real mouth-to-mouth kiss is shared a year or so later, and by the age of eighteen the majority of teens will have had sex.

The link between rising sex hormones and sexual interest and capacity shouldn't come as a surprise, because this is what puberty is all about—preparing our bodies and behavior for reproduction, which involves mate seeking, courtship, sex, and care of children. And that's what the sex steroids do. They affect gene transcription to coordinate large and often relatively slow changes in multiple body systems, including the brain, and have long-lasting effects. And it's not the small changes in the sex hormones that stimulate sexual desires and capacities, it's the big ones—like the striking increases across puberty and declines that accompany old age.

In girls, T levels barely rise from the almost undetectable levels in childhood to low levels in adulthood, but progesterone and estrogen show pronounced increases. In female mammals (including humans), the effect of the changes in hormones on sexual behavior is clear. The females become the focus of male (and sometimes female) sexual desire, and both sexes get pretty excited about the possibility of getting it on. In boys, T begins its rise at around ages nine to ten, then increases more sharply from thirteen to fifteen, leveling off around age seventeen, plateauing, then declining slowly, especially after age forty.

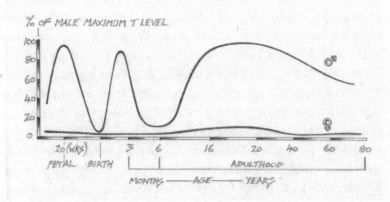

% OF MALE MAXIMUM T LEVEL

T levels across the life span in males and females

As the figure above shows (and as we saw in chapter 3), way back in the womb is where the sex difference in T starts—boys are exposed to much higher levels than girls, at as early as eight weeks of gestation. This higher T masculinizes the brain (organizational effects), as I explained in chapter 4. It's during puberty that these previously masculinized brain structures are again exposed to high levels of T, which further shapes them in response to the individual's experiences and environment, and activates them for male sexual behavior.

Given the evolutionary pressures on each sex to maximize reproductive success, how would we expect men and women to differ in their sexual desires? And given what we know about the importance of the environment, how would we expect men and women to differ in their actual behavior?

EVOLUTION, ONE-NIGHT STANDS, AND STAYING HOME

I can't say it enough: culture matters. Here are just a few examples of the many ways in which the social environment can affect sexual behavior. Boys of the Sambia people of Papua New Guinea perform oral sex on men, with the ingested semen said to make them into men in their turn; we can only imagine what would happen if someone suggested importing the practice. Mormons in the United States historically practiced polygamy, with their founder Joseph Smith enjoying the benefits of around forty wives; the Edmunds Anti-Polygamy Act became law in the United States in 1882. In ancient Rome, homosexual behavior was thought to be masculine and acceptable for the men who did the penetrating; to be penetrated, however, was considered a submissive, servile act. In the modern world, in much of Africa and the Middle East, homosexuality is illegal and punishable by death, whereas in much of the Americas and Europe, homosexual marriages or unions are legal. In many parts of the United States, young women are likely to be "slut-shamed" for having a relatively high number of sexual partners (women from high socioeconomic backgrounds are more likely than other women to be so shamed). On the other hand, promiscuous young men may be praised as studs (men from low socioeconomic backgrounds are likely to get more praise than other men for such behavior).

Even with all of this variation, an evolutionary perspective suggests that there should be a biological signal through all this cultural noise. We would expect men to be more open to casual sex than women, prefer a higher number of sexual partners, and have a stronger sex drive (libido). On the other hand, both sexes should be more or less equally keen to form long-term partnerships.

In theory, men's reproductive success is limited only by the number of one-night stands they can get—and have the stamina for. But the evidence suggests that the most successful men in most hunter-gatherer societies sire only around twenty-five children (a figure that is not completely out of reach for exceptionally fertile women). The prize for superlative achievement in the fatherhood stakes goes to Genghis Khan, the thirteenth-century emperor of the Mongol Empire, who probably had hundreds of children, easily surpassing what is possible for women. Genghis's sons inherited his appetite for promiscuity, and around one in two hundred of all men alive today have the First Great Khan as an ancestor. Exceptional male fecundity of this kind is harder to achieve in today's societies but hasn't completely ended. An Angolan man aptly nicknamed "Big Dad" recently died, reportedly leaving behind 156 surviving kids and 250 grandkids.

Most human fathers are nowhere near as prolific as Big Dad and Genghis Khan. As usual, sex differences—this time in reproductive success—are most pronounced at the extremes. Unlike red deer stags, men don't usually have harems. Part of the reason that a stag can get away with having a harem is that his kids can survive without him. Like many grazing animals, which face the threat of being somebody's lunch if they can't run away, newborn red deer calves can get up and walk within hours after birth. On the other end of the spectrum are song sparrow chicks, born totally helpless and completely dependent on care from both parents. They are blind, can't fly, and need mom to keep them warm. Without dad's care, they are far less likely to survive. If dad doesn't want his genes to stop with him, he needs to stick around. This helps to explain why pair bonding for a season or more—even for life—is the preferred mating pattern in about 90 percent of bird species.

Human babies are a bit of a mix—they can see and keep themselves warm, but they can't get around on two legs for about a year, and it takes even longer than that for them to be able to find their own food. They can, of course, survive without dad, but in many societies today, and over evolutionary history, a human baby's chances are much better if he stays to help out.

A woman living in a hunter-gatherer society can't rely on Whole Foods to deliver the calories needed to nurse and care for her young. The maximum rate at which she can produce children is around one every three years, which is about eight kids over a reproductive lifetime of twenty-five years, assuming no twins or triplets. For a woman leading a lifestyle typical of our ancestors, securing one high-quality, high-status, high-investing mate was usually a better reproductive bet than increasing her number of sexual partners. If the father can help the mother to be healthier, reduce her stress, and take in more calories, her offspring have a better chance at survival. And if she has more energy available to devote to reproduction, this may also decrease the time between births, allowing her to produce more babies over her lifetime. While other family members can assist, husbands are especially important in helping mom to stay in prime reproductive condition. Not surprisingly, women tend to prefer mates who don't just have high social status but who also signal willingness to invest in them and their children. Men who fail to signal these qualities and who are just out for a good time may have a harder time finding healthy, fertile mates because women are exercising their reproductive choices.

Sticking around to help out with one's own mate and children, for most men, is likely to be a lower-risk, higher-payoff reproductive strategy than competing on the open market for new mates. Considering all human societies for which we have reliable records, a monogamous partnership is the typical form of mating arrangement. But having said that, men generally benefit more reproductively than women do from increasing their number of mates, and polygyny (many wives) is much more common than polyandry (many husbands).

From an evolutionary point of view, the best way for humans to maximize the survival of their offspring is for both parents to stay home; that is, to work as a team. Parents in the distant past relied on lots of help from extended family or other community members. But today, many parents lack this kind of support. Even without it, evolution encourages parents of all kinds—including adoptive or single parents, two moms or two dads—to stay in the game. We come equipped with a set of psychological adaptations that help us cope with the high demands of raising children. For those with a partner, romantic love solidifies bonds and can make the teamwork tolerable. And parental love can motivate involvement and sacrifice, and even make the whole enterprise feel rewarding—on balance, anyway!

Still, investing heavily in offspring, both during and after pregnancy, is compulsory for women in a way that it is not for men. Especially when it comes to what researchers call "short-term matings" (and everyone else calls "one-night stands"), men who are motivated to make and seize the opportunity should do better at bequeathing their genes to subsequent generations. We should expect the mating strategies of men and women to significantly overlap, especially when it comes to pair bonding. But when it comes to eagerness for sex and taste for variety, we should expect a larger separation. And that is exactly what we find.

SEX DIFFERENCES IN BED AND IN THE LAB

Unless you're living under a rock, you might suspect that stereotypes about men being hornier than women and wanting more sex without strings have some truth to them. And they do. But perhaps that is some culture-bound contingent truth, overturned in other places. Fortunately, researchers have collected a huge amount of data on sex differences in libido and preference for sexual novelty, using a variety of different methods, and from a wide range of diverse cultures.

For example, in a 2009 study, psychologist Richard Lippa and his team analyzed data from a massive internet-based survey conducted by the BBC in collaboration with scholars from all over the world. The use of the internet allowed a huge number of responses. Two hundred thousand people responded, from fifty-three nations, including the United States, Pakistan, Brazil, Russia, India, Singapore, and China. (Most were from the UK and the United States.) Taking the survey responses at face value, the researchers showed that, on average, men have a significantly higher libido than women, even if they can't always act on their sex drive with an actual partner.

To measure desire for sexual novelty, the researchers asked some questions from the Sociosexual Orientation Inventory, known as the "SOI." This is a widely used instrument that assesses people's willingness to have casual sex. Remember the character Samantha, from the popular HBO series *Sex and the City*, from the late 1990s? She was sexually liberated, to put it mildly. Many episodes featured Samantha's sexual enthusiasm for her latest conquest (and she had many). She would have been at the top of her sociosexuality class, achieving one of the highest possible scores. (If you're interested in seeing how you stack up, you can Google "sociosexuality inventory" and take the test yourself.) The shortened version used in this study comprised these three questions, answered on a 7-point scale, ranging from "disagree" to "agree": "Sex without love is OK," "I can imagine myself being comfortable and enjoying casual sex with multiple partners," and "I would have to be closely attached (emotionally and psychologically) to someone before I'd feel comfortable and fully enjoy having sex with him/her."

Unsurprisingly, men scored higher than women. This greater desire for sex without commitment among men is one of the largest of all psychological sex differences in humans. To give you a better idea of the size of the difference in sociosexuality averaged across all of the countries, if you were to select a woman at random, there would be a 70 percent chance that she would be less interested in casual sex than a male picked at random. The size of the sex difference in libido across all countries is slightly smaller.

This pattern was found in every single nation—without exception, men scored higher in sociosexuality. But the *size* of the sex gap varied a lot by country. Wealthier and more gender-equitable countries like Iceland, Austria, Denmark, Sweden, and France had smaller gaps, mostly because the women in those countries expressed more sexual freedom (higher sociosexuality), and their scores were closer to those of the men. On the opposite end of the range were the less gender-equitable countries with the larger gaps in sociosexuality, such as the Philippines, Pakistan, Saudi Arabia, and Turkey. In these countries, women's sexual behavior was relatively restricted (lower sociosexuality), resulting in a larger distance from the men's much higher scores. (The sizes of the sex differences in sociosexuality in the United States and the UK were in the middle of the range for all the countries.) But the sex difference never reversed, and Lippa concluded that the data supported the theory that sociosexuality results from "cultural influences superimposed on biological predispositions."

In contrast, although men everywhere reported a higher sex drive, as measured by the responses to, "It doesn't take much to get me sexually excited," and "I have a strong sex drive," the strength of people's libido and the size of the sex gap didn't vary much by country.

That libido is less culturally malleable shouldn't be surprising. Our environment strongly affects what we believe and how we act, but its impact on our natural drives, like those for food, sleep, and sex, is far weaker. The motivation to eat is likely to be just as strong in Pakistan as in Denmark, even though the expression of that drive is quite different. In Denmark, pork is a popular and much-loved meat. But no law-abiding Pakistanis eat pork, which is forbidden under Islam. Danish cultural values may lead Danes to believe that pork is healthy and good, and Pakistanis may believe that it is unhealthy and disgusting. Even if sex drive is relatively immune to cultural influences, it is easy to imagine how social standards and expectations might push around answers to the survey's questions about sociosexuality, like "Sex without love is OK."

THE PENIS NEVER LIES

These results are the tip of an iceberg of evidence from other large, cross-cultural studies that point in the same direction: men have a higher libido than women and a greater preference for sexual variety, with the latter difference being more strongly subject to cultural influences. Of course there are problems inherent in surveys that rely on self-report—respondents may not be answering honestly, for one thing. These problems can be overcome to some extent by including questions designed to ensure that answers are consistent. And then there are other, more creative methods, like hooking people up to fake lie detectors.

But the penis never lies. At least, that's what sex researchers seem to think, and it strikes me as a pretty solid observation. We can't stick microdialysis probes into the brains of men and women as they get it on or ask them to switch out familiar sex partners for new ones just to see what happens. But we can let sex researchers use some of their favorite tools, and the penile plethysmograph has got to be near the top of the list. It involves strapping a ring onto a man's member that registers changes in its circumference as he watches nature documentaries or pornography, or listens to erotic recordings or classical music.

When men—mostly college undergraduates in Europe and the Americas—come into the lab for one of these experiments that include more "objective" measures of arousal like those from the plethysmograph, the results match their self-reports. In other words, what the penis says about how turned-on a man is matches his verbal reports about how turned-on he is.

As a man watches pornography, his sexual arousal (as indicated by changes in penile tumescence) wanes over time if the actors in the movie don't change. But it pops right back up when new actors come on the scene. One group of researchers included other measures of participants' arousal in response to watching pornography: the amount of time it takes them to ejaculate, the volume of that ejaculate, and the hardiness of the sperm (the portion that were good swimmers). All of these measures show that, like in the rat, a hearty sexual response returns in men who have sexual access—or the virtual suggestion of sexual access—to a new mate.

The converging evidence for the Coolidge effect in men—regardless of sexual orientation—is strong. If the effect is present in women, it is much weaker.

Other lines of evidence support this sex difference in the preference for novelty and in libido. For instance, men out-masturbate women. (They also make more stupid mistakes about where and when they do it. The writer and CNN commentator Jeffrey Toobin recently provided me with an example of this when he neglected to ensure his camera was off while he pleasured himself during a Zoom conference with fellow *New Yorker* staff.) Men are the main consumers of pornography worldwide, which allows them to view naked, apparently sexually eager women (or men) without working for it. (And men prefer to dispense with the "emotional foreplay" that women like in their pornography.) They fantasize more about sex with strangers and multiple partners, while women tend to fantasize about current partners or people they know. Men are almost exclusively clients of prostitutes, the world's oldest profession. They are more likely than women to use infidelity websites and to respond positively to invitations to have sex with someone they don't know.

MEN AND SEXUALITY: IS T TO BLAME?

Given what we know about human evolution, and the evidence for competition between men for sexual access to women, the most parsimonious explanation of these sex differences is that men are adapted to have a greater sex drive than women and a stronger preference for sexual novelty. Culture matters, but there is no evidence that it can account for the consistent patterns of sex differences in sexuality across the globe.

Adaptations need mechanisms, which arise from the genes that have been preferentially selected because they confer a reproductive advantage. Our new housecat, Lola, is adapted for hunting. She loves to hide behind the living room sofa, stalking a small, fur-covered ball, and then dart out, pouncing on her "prey" and attempting to use her sharp claws and large canines to disembowel it. Her claws and canines are part of a complex heritable mechanism that allows her to hunt effectively.

Once it is granted that men's greater sex drive and preference for novelty are an adaptation, there is little doubt that testosterone is part of the mechanism. Whatever the mechanism is, it of course has to differ between males and females. And high testosterone, a product of male sperm-producing testicles, clearly promotes physical and behavioral features designed to increase mating success. There is every reason to think that the mechanisms that explain the greater male libido and preference for sexual novelty likewise involve T.

Animal studies aside, we know that large increases in men's T levels, going from extremely low to normal, will increase sex drive, sexual arousal, and sexual function. And the reverse is true. As you'll see in the next chapter, those measures all decline for men who shut down their T production for medical reasons, or as part of a gender transition.

A man's T responds to physiological, social, and environmental circumstances to incline him to behave in ways that have been, on average, reproductively beneficial. Sometimes that will involve seeking out multiple partners, but often it won't. Remember the male song sparrows from chapter 6, whose T levels dropped when they became dads? I mentioned that something similar also happens in men. First, T may drop in newly paired men who are in committed relationships, and lower T in this situation is associated with being a more devoted partner. Conversely, higher T is more characteristic of men who are keeping an eye out for opportunities to cheat.

When the baby comes, T may drop even more. This drop in T, along with other hormonal changes, may serve to make the hard work of parenting feel less like a chore and more like a rewarding and enjoyable activity (the balance there is, of course, often tipped in the other direction). In addition, like the low-T dad sparrows, it may dampen one's drive to compete with other males and seek out other mates.

A man's hormonal response to becoming a father depends heavily on his culture and the amount of time he interacts with his children. For example, dads in the Hadza foragers of Tanzania (who are generally socially monogamous) frequently hold, feed, and play with their babies, while dads in the neighboring Datoga pastoralists (who practice polygyny) are more likely to leave this to the moms and other caregivers. You can guess which dads have the lowest T levels—the Hadza. The T levels of the Hadza dads were found to be almost 50 percent lower than their childless fellows, while the T levels of the Datoga dads were no different from Datoga men without kids.

Physically interacting with young children, doing things like feeding, playing, carrying, or changing diapers, is associated with a T drop in dads. And in many situations, a dad's focus on his family will increase his reproductive success. (Lower T in dads is not *always* better, though—T has been shown to rise in response to cries, perhaps to fuel a drive to protect.)

DOES T TURN WOMEN ON?

By this point in the book, you might be wondering what T does in women, apart from serving as a precursor to estrogen. The word "estrus" is derived from the Latin *oestrus*, meaning "frenzy." Female animals do sometimes appear to be in a frenzy when they are seeking out sex—women included!—but the evidence implicating T is weak.

Perhaps unsurprisingly, there is no lack of research on the role of T in male sexuality—libido, sexual performance, erectile quality, etc. We know far less about the role of T and other sex hormones in influencing women's sexuality. One reason for this is a relative lack of research, which could be related to the complexities of women's hormonal fluctuations. Sex hormones including T change considerably across a woman's menstrual cycle. For example, estrogen and testosterone both peak around ovulation, and a woman's T level at the start of her cycle will be very different from that at the middle of it. And as I explained in chapter 5, accurate measurement of women's T is not possible with the traditional technique, radioimmunoassay. So if no relationship is detected between T and aspects of female sexuality, like sexual desire and function, that may be due to methodological problems, or there may really be no relationship.

If the idea of sex—either with a partner or solo—holds little appeal, or if sexy thoughts are few and far between, you might have "low sexual desire." This is only a problem if it bothers you. Lots of factors affect your interest in sex, like your age, mental and physical health, and whether you're in a relationship.

Across the globe, about twice as many women as men report that they aren't particularly interested in sex, but the rates of low libido vary widely across places and ages. For example, in one of the largest studies of its kind (surveying people from twenty-nine countries, including the United Kingdom, Germany, the United States, Australia, Turkey, Indonesia, and South Africa), researchers asked about 20,000 men and women, who ranged in age from 40 to 80, about their desire for sex. Low libido in women was most common in the Middle East and Southeast Asia (43 percent), and it was least common in women from Northern Europe and Central and South America (26 and 29 percent, respectively). Low libido is less common among younger women, reported by 20–30 percent of women under 40 in the United States.

Doctors frequently prescribe testosterone "off-label" (i.e., not approved by the U.S. Food and Drug Administration) to women who seek to increase their libido. However, there's little reason to think it works. (And men may also be overprescribed testosterone for declining libido with age; as in women, the causes are complex.)

Most women of reproductive age (late teens to late thirties, with wide variation) who complain of low libido do not have low T levels. Supplementing these women with additional T has no effect on sexual desire or behavior, unless T levels are raised far above the normal female range. In that case, libido does tend to increase, but so, too, do the masculinizing effects of T on the body, like acne and facial hair. (As we'll see in the next chapter, for some females seeking to transition to male, T-induced masculinization is far from undesirable—it's the goal.)

Postmenopausal women (generally women in their midfifties and older), whose ovaries have for the most part closed up shop, are more likely to complain of low libido than women of reproductive age, whose ovaries are still open for sex-steroid-producing business. The low levels of androgens that are in circulation after menopause come mainly from the adrenal glands, and this is also the primary source of women's estrogen postmenopause. T levels drop by 30 to 50 percent after menopause, and since T is so clearly associated with sex drive and function in men, it's become a prime suspect in the decline in these same parameters in postmenopausal women.

As always with female sexuality, the situation is complicated. Low libido (in men or women) may or may not have a hormonal cause. We age along with our partners, and our energy and health may move in opposite directions from our age. Some studies on the efficacy of T to increase libido in postmenopausal women report a small effect, and others no effect. Overall, the findings on T and libido in postmenopausal women are weak and inconclusive.

The ultimate test case for whether T is crucial for typical female sexuality would be someone without any T action whatsoever; if she had little or no sex drive, or problems with sexual function, then this would indicate that the hormone is essential for normal female sexuality. However, people with complete androgen insensitivity syndrome (CAIS—remember Jenny from chapter 3), who are entirely immune to the effects of androgens but who experience all the effects of estrogen, appear to have typical sexual responses, desires, and orgasmic ability.

Sexuality is obviously complex, and perhaps more complex for women. Social and emotional factors play a more important role in sexual motivation and arousal in women than in men. Dopamine is upregulated in women in sexual contexts, as it is in men, but this occurs primarily with the backdrop of sex hormone levels that are typical of women: relatively high estrogen and low testosterone. And we know from studies on both nonhuman females and women that estrogen is essential to sexual motivation. As with many female behaviors, we women have our own ways of doing things.

So let's not fall into the trap of thinking that because T is essential for male but not female sexuality, men are sex fiends and women lie back and think of England. If you go back to the endocrinologists' joke about the Coolidge effect, you'll see that it has an ironic touch: it's Mrs. Coolidge who is more interested in frequent sex than her husband.

GAY RATS

I relied on the Coolidge effect in rats to introduce the topic of T and men's preference for novelty. The humble rat can also introduce the second topic of this chapter—T and sexual orientation.

As far as we know, there aren't any gay rats in nature—that is, rats exclusively sexually attracted to their own sex. (In fact, the only clear example of exclusive homosexuality in a nonhuman animal is the male domestic sheep.) But nonexclusive homosexual behavior is common among animals, including rats. As I mentioned in chapter 4, sometimes female rats will try to mount other rats (of both sexes), and some males will even adopt the female lordosis pose in response. And researchers know how to increase the frequency of these behaviors—by manipulating the rats' endocrine profile. Removing the gonads and supplying cross-sex hormones can induce such homosexual behavior.

The way to do it is to alter a rat's exposure to T during its sensitive period, which is not prenatal but rather in the first few days after birth. We can create a female rat who prefers to mount other females simply by giving her the T treatment that normal males get: high levels of T during that early sensitive period, and again in adulthood. And we can do the same for males, by castrating them and mimicking the female sex hormone environment in adulthood. Similar effects of sex hormones have been found in hundreds of studies on many other animals. Might human same-sex attraction also have something to do with unusual patterns of sex hormone exposure?

LESBIAN MECHANICS AND GAY FLIGHT ATTENDANTS

Rats give us some hints about the role of hormones in human sexual orientation, but human homosexuality is far more complex and interesting. In humans, patterns of sexual attraction are associated with more than just the target of sexual attraction. For example, there's the stereotype reflected in the heading above that lesbians are more likely than straight women to be mechanics, and gay men are more likely than straight men to be flight attendants. It's just that, a stereotype, and of course it doesn't reflect the full diversity of career preferences in lesbians and gay men. But in this case, the stereotype is accurate.

Compared to heterosexual women, lesbians are more likely to be attracted to male-dominated occupations—such as truck driving, building contracting, and appliance repair—that have more to do with things than people. And gay men are overrepresented in female-dominated occupations—such as hairdressing, nursing, and interior design—that have more to do with people than things. In other words, relative to heterosexual people, homosexual people are more interested in careers that are typical of the opposite sex. And perhaps unsurprisingly, homosexual people who show such preferences rate themselves as less masculine or feminine than their same-sex heterosexual counterparts. These gender-atypical feelings don't just pop up in adulthood; for those who have them, they were there from the beginning.

A gender-atypical trait in males is simply one that is more characteristic of females in a given population, and vice versa for a gender-atypical trait in females. ("Gender-atypical" implies nothing about origins or whether the trait is good or bad.) For example, when I was in grade school I played Little League baseball. This was a "gender-atypical" behavior because in 1970s New England, Little Leaguers were almost all boys. (But times have changed, and plenty of girls play Little League baseball, so the behavior is no longer gender-atypical for girls.)

Boys who show gender-atypical interests—shying away from rough-and-tumble play, gravitating toward dress-up or makeup, playing with dolls and girls—are more likely to grow up to be gay than are gender-typical boys. Similarly, girls who shun dresses and prefer rougher sports and horseplay with the boys are more likely to grow up to be lesbians. This association between gender-atypical interests in childhood and later homosexuality or bisexuality has been found in diverse cultures, from the United States to the Philippines, Samoa, Guatemala, the UK, and Brazil.

Here's an example of the relevant research. In a classic study, one of the largest of its kind, Richard Green (then a psychiatrist at UCLA) and his team followed a group of boys from childhood into early adulthood to determine whether early gender-atypical behavior predicted adult homosexuality. They recruited sixty-six boys between four and eleven years old, who showed "extensive cross-gender behavior." This group of "feminine" boys was compared to another group of fifty-six boys, demographically matched to the first group but not selected on the basis of their gendered behavior. Most—70 percent—of the gender-atypical boys frequently dressed up in clothes that were typical of girls. Not one of the gender-typical boys did this frequently, although 20 percent did so rarely. Even more of the gender-atypical boys—85 percent—stated that they wished that they were girls, while only 10 percent of the boys in the gender-typical group said that they felt this way.

Years later the researchers interviewed the boys, now adults or in late adolescence (some had dropped out of the study along the way). All of the men in the gender-typical group were heterosexual, and 75 percent of the men in the gender-atypical group were gay or bisexual.

Were these little boys socialized to be gender-atypical? There was no evidence for that. On the contrary, their behavior persisted in the face of strong social discouragement. Many gender-nonconforming children have felt the harsh sting of disapproval and shame, including rejection from peers and family.

By now you're probably thinking of the obvious hypothesis. The feminine boys had low T exposure in the womb, and this affected both their childhood behavior and later sexuality. And for masculine girls ("tomboys"), it's the other way around—they had high T exposure in the womb, which inclined them toward gender-atypical play as children and a same-sex orientation in adulthood.

How could prenatal T influence gender-atypical play? It could nudge the child to have a preference for the behavior itself, or its effects could be less direct. T could act to reduce fear, anxiety, and increase risk taking and a taste for novelty, or it could nudge the child to have a preference for behaviors typical of the opposite sex, whatever they might be (or all of the above). For example, if I had high T in utero, it might have directly increased my desire to hit and throw things, or it might have inclined me to want to play like the boys I knew. It's hard to know which is right. But a direct influence is more likely when T is also associated with a particular behavior in nonhuman animals, as is the case with rough-and-tumble play or when we see consistent patterns across times and cultures in the relationship between T and behavior. It's plausible that both direct and indirect actions of T are at work, particularly in adults.

So does higher T prenatally incline us to become masculine and attracted to females, and lower T incline us to become feminine and attracted to males? Well, it's complicated. For women, there's some evidence for a role of T in sexual orientation, but in men, the evidence is weak or nonexistent.

FETAL T AND SAME-SEX ATTRACTION

In chapter 4, I described the behavior of girls with congenital adrenal hyperplasia (CAH). This is a condition in which females are exposed to higher-than-usual T levels in the womb. These girls are more likely to play in the rough-and-tumble boy style than their nonaffected peers, and as adults they are overrepresented in male-typical occupations. CAH girls grow into women who are far more likely than other women to be lesbians—about 30 percent of girls with CAH grow up with sexual orientations characterized as "not exclusively heterosexual." That is still a minority but is much higher than the prevalence in the general population (4 percent or so).

We can also consider an opposite kind of case, CAIS girls. (Recall, again, Jenny from chapter 3.) They have male-typical XY sex chromosomes but cannot experience the effects of androgens at all. They are no different from typical XX girls in their play preferences, and they grow up to be almost exclusively attracted to men.

However, these are extreme examples. Most people develop with levels of T within the normal range for their sex, in utero and beyond. So—one might think—most gay men just got a little less of the molecule of masculinity than heterosexual men, and gay women got a little more than heterosexual women. How could we find out?

What we want to know is the relative level of testosterone to which male and female fetuses are exposed—are they on the high or low end for their sex? But T timing is a big issue, too. Prenatal T is high in male fetuses between weeks 8 and 18 of pregnancy. The genitalia differentiate during the first part of that period, and evidence suggests that the brain differentiates during the second part. To complicate matters, T also rises in males during "mini puberty," in the few months after birth, and this might also affect sexual orientation. (See the figure on page 189 above.)

If T affects sexual orientation pre- or directly postnatally, we'd want to measure the amount of T in the fetal blood during the critical period, not at other times. Unfortunately, accomplishing all of this is pretty difficult—in humans, anyway. Pregnant moms don't want to get poked and prodded just to benefit science; they want to do it when necessary to ensure the health of their baby. T is typically measured from samples from amniotic fluid or umbilical cord blood, and these fluids may not accurately reflect the T levels to which the fetus is exposed. Moreover, we do not know the exact timing of the crit-

ical periods for brain differentiation in humans, so we can't be sure
that we're measuring T at the right time. Given these methodological
problems, perhaps it's not surprising that the results of such measurements don't give us much insight into the hormonal origins of homosexuality.

CAN WE GET YOUR DIGITS?

Another popular method for estimating T levels in utero is one's "digit
ratio," which represents the ratio of the ring finger to the index finger. I know, sounds weird. But hold your hands up, palms facing you,
with your fingers straight. You might notice that the relative lengths
of the ring and index fingers are different on each hand. If you're a
woman, your index finger is probably the same size as or slightly longer than your ring finger, and the reverse if you're a man. (See the figure below.) This is a sex difference that holds across many vertebrates,
is present in utero, and can be manipulated by altering T levels. As
a group, girls with CAH (who had high prenatal T) have lower digit
ratios (meaning shorter index relative to the ring finger) than unaf-

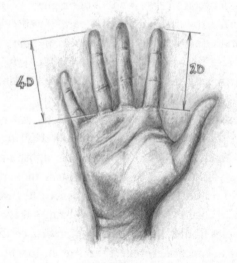

Measuring digit ratio

fected peers. A low ratio is the masculine profile, and a higher one is the feminine profile.

The digit ratio is a "noisy" signal of early testosterone—many things other than T affect it. To be confident we're seeing the effects of prenatal T on any particular trait, we need to sample hundreds of people. Even though digit ratio is a blunt tool to investigate prenatal T and doesn't tell us much about any individual, the technique could potentially uncover meaningful group differences. And since measuring one's digit ratio is a relatively easy, cheap way of relating someone's hormones to a behavioral trait, lots of studies have been published that relate it to all kinds of outcomes—including aggressiveness, cognition, athleticism, sexual behavior, and, of course, sexual orientation. Some of these make a big splash in the media, and most of them report significant results (one paper even claimed that digit ratio predicted belief in the paranormal!). However, this could be because the experiments that fail to find an effect are less likely to be published than experiments that do.

With that caveat in mind, what do the digit ratio studies show? Some find that lesbians tend to have a lower (more masculine) ratio—but others fail to find any such effect. As far as gay men are concerned, there's little evidence from digit ratio studies that they had lower T prenatally, and, like lesbians, their adult T levels are no different from those of their heterosexual counterparts. What's more, to throw a large penis-wrench into the low-T theory, one study reported that gay men have bigger penises than straight men (6.3 versus 6 inches when erect), and penis size is positively related to early T exposure.

Testosterone has well-documented effects on sexually dimorphic anatomy and behavior, especially on aggression and sexuality. But we have a long way to go in understanding the role of T, if any, in the origins of sexual orientation. Although the evidence for a link in women is stronger than in men, as suggested by the higher prevalence of gender-atypical preferences and interests and lesbianism in CAH females, T could affect sexual orientation in males during prenatal development in a number of different ways that are presently hard to detect.

GAY LESSONS

Although the relationship between T and male sexual orientation is still up for debate, gay men have something to teach us about the first topic of this chapter—male sexual behavior—and in particular the preference for sexual novelty.

In an article in *New York* magazine titled "#MeToo and the Taboo Topic of Nature," the prolific writer and social commentator Andrew Sullivan told an old joke: "What does a lesbian bring on a second date? A U-Haul. What does a gay man bring on a second date? What second date?"

Like the Coolidge effect story, this joke plays on stereotypes, this time about the sexual and romantic proclivities of homosexual men and women. And again, there is truth in the stereotypes. In comparison with their straight cousins, on average, gay men do have many more sexual partners. Lesbians do not, and are much more likely to be sexual within committed monogamous relationships.

There is little evidence that gay people are gender-atypical in terms of libido, the desire for casual sex, preferences for age, and the importance of good looks in potential mates. Even gay men who had feminized play styles as kids do not have a feminized pattern of sexual behavior in adulthood, apart from the target of their sexual attraction. Lesbians and heterosexual women, relative to gay and heterosexual men, are more likely to be in a monogamous relationship, are less interested in casual sex, and want fewer sexual partners over their lifetime. That is, regardless of sexual orientation, adults with male levels of T tend to have the sexual proclivities of men, and those with female levels tend to have those of women.

The sexual interests and behavior of people in same-sex relationships provide insight into how each sex might behave if freed from the constraints and expectations of the opposite sex. As a gay man, Sullivan has some personal experience and speculates on whether culture or biology is at the root of these patterns:

Is male sexual aggression and horniness a function of patriarchy or testosterone? ... My suspicion is that it's more about nature than about society, and one reason I believe this (apart from all the data) is because I'm gay. I live in a sexual and romantic world without women, where no patriarchy could definitionally exist, a subculture with hookups and relationships and marriages and every conceivable form of sexual desire that straight men and women experience as well. ... In fact, remove women, and you see male sexuality unleashed more fully, as men would naturally express it, if they could get away with it. It's full of handsiness and groping and objectification and lust and aggression and passion and the ruthless pursuit of yet another conquest.

This does not mean that gay men only want to have lots of anonymous sex, or that it is the type of sex that all gay men want. Of course, gay men have long-term relationships, including marriages full of emotional intimacy, love, and commitment, and lesbians can stay single for life. The important point is the connection Sullivan sees between the sexual attitudes and behavior of gay men, and men in general. And here, I think he is right.

Of course, gay culture is quite hospitable to men having multiple sex partners, but it's not likely that culture explains why gay men are relatively promiscuous. The wider straight culture doesn't encourage gay male promiscuity—generally quite the opposite. And there's nothing to stop lesbians from having their own bathhouses and anonymous hookups, but they don't. Gay, lesbian, and the less-studied bisexual cultures seem to follow the inclinations of gays, lesbians, and bisexuals, not the other way around.

The obvious explanation is that men are more motivated to have sex and have a stronger preference for a variety of partners. Gay men have more sex simply because they can: it's not a "gay" thing, it's a "man" thing.

T in Transition

T TRANSFORMATIONS

National Public Radio's *This American Life* once devoted an entire episode to testosterone. In one segment, the producer interviewed Griffin Hansbury, a female-to-male (FtM) transgender man. Here is a little of what he had to say about how he felt when his newly elevated levels of testosterone started to kick in early in his hormonal transition:

> I remember walking up Fifth Avenue, and there was a woman walking in front of me. And she was wearing this little skirt and this little top. And I was looking at her ass. And I kept saying to myself, don't look at it. Don't look at it. And I kept looking at it. And I walked past her. And this voice in my head kept saying, turn around to look at her breasts. Turn around. Turn around. Turn around. And my feminist, female background kept saying, don't you dare, you pig. Don't turn around. And I fought myself for a whole block, and then I turned around and checked her out. And before, it was cool. When I would do a poetry reading, I would get up and I would read these poems about women on the street. And I was a butch dyke, and that was

very cutting edge. And that was very sexy and raw. And now I'm just
a jerk.

It was like being in a pornographic movie house in my mind. And
I couldn't turn it off. I could not turn it off. Everything I looked at,
everything I touched turned to sex.

Hansbury is one of many people who have transitioned with
cross-sex hormones (also called "gender-affirming" hormones). People
vary in their responses to hormone therapy, and those responses also
change over time. For example, when Hansbury first started taking
T, he was in the midst of something like male puberty, which can feel
overwhelming (as he described above) at first. That being said, people
undergo such therapy because they want the end result: the predictable changes that it delivers to the body and mind.

Trans people who drastically alter their T levels are in a unique
position to offer insights into how life is different when they cross
over to the other side of the testosterone line. In this chapter a few of
them will tell what it is like. What they describe is consistent with the
broader literature on T in gender transitions.

This chapter is about their experiences on and off of testosterone.
Changing T changes beards, brains, Adam's apples, and so much
more. As we'll see, for those wishing to physically transition from one
sex to the other, T is sometimes part of the problem and sometimes
part of the solution.

WHY TRANSITION?

"Transgender" is a broad term, referring to "the full range of people
whose gender identity and/or gender role do not conform to what is
typically associated with their sex assigned at birth." The number of
people who identify as transgender has been rising quickly, although
the reasons for this are unclear. A recent literature review estimates
that in the United States in 2017 1 in 250 (about one million) people
identified as trans, double the figure a decade earlier. (Other estimates
give higher numbers.) Identifying as nonbinary—neither male nor

female—has become much more popular, and there are now more young people than older people with transgender identities.

Many trans people suffer, or have suffered, from "gender dysphoria"— distress and anxiety about features of one's sexed body and the gendered ways in which one is perceived by others.

It might be hard to imagine how that feels. As an imperfect analogy, consider how discomfort with various physical traits could affect your own mental health. Many of us can relate to the feeling of shame about our body. We may feel that we have too much fat, are too short or too tall, or have skin that seems too old and wrinkly. We might feel that our breasts are too small, that we have too much facial hair, or that our biceps are too small or voice too high. We may feel that such features prevent us from being seen for who we truly are or prevent us from being seen at all. Some people feel so self-conscious about their bodies that they adopt unhealthy eating or exercise habits, or become anxious, isolated, lonely, and depressed. They would do anything to change that. Across the globe, the "aesthetic medicine" market is valued at many billions of dollars.

The transgender celebrity Jazz Jennings says that she knew she was a girl as soon as she could talk, and she was diagnosed with gender dysphoria at the age of three. Gender dysphoria can also begin later, in adolescence or adulthood. In children, the distress usually resolves by the end of puberty, and even if it doesn't, some adults manage to cope. But for those with acute dysphoria that shows no signs of abating, transitioning can bring relief. This might mean going all the way to the other sex, or ending up somewhere in between. Whatever the end goal, embodying one's desired identity, as far as that is possible, can also bring joy and a new sense of comfort and freedom. In any event, the transition will involve social changes—clothes, hairstyle, pronouns, and so on. Often the transition will also be medical: therapy with cross-sex hormones, and perhaps surgery on breasts, genitals, and even facial bones. Jazz's transition, which included the fashioning of a neovagina from her undeveloped male genitalia, was covered in detail in the TLC series *I Am Jazz*.

Recently the number of children and adolescents referred to gender clinics has risen sharply, with girls significantly outnumbering boys, and a consequent rise in treatment with T. In England, for example, the number of young natal females (children born female) referred to the National Health Service's Gender Identity Development Service has increased fifty-fold over the last decade, a pattern seen in other countries. Hormones and surgery for transgender people are booming.

The effects of male levels of testosterone when taken by natal females can be spectacular. This is because both sexes have androgen receptors (the gene that codes for the receptor is on the X chromosome, and females have two of those); females thus have the capacity to respond to high levels of testosterone. In fact, when testosterone is high, it helps to make more of its own receptors.

Google "Buck Angel" and you'll see what I mean: Buck is a cigar-smoking, muscular guy with a full beard (sometimes he sports a goatee), rather like a hairier and heavily tattooed version of the action hero Vin Diesel. But, as Buck's website explains, "Buck was born female on June 5, 1962. He never felt female and struggled through life until he had the life-changing opportunity to transition from female to male and finally live life authentically." It wasn't until Buck was twenty-eight that he started taking T. His voice dropped, his clitoris increased a few centimeters, he grew a beard, and, by hitting the weights, he packed on a huge amount of muscle.

ALAN'S STORY

To learn firsthand what it's like crossing to the other side of the T tracks, I talked to Alan—a trans man, like Buck and Griffin Hansbury:

> As a little kid, I was definitely a tomboy. My friends were mostly boys, and I was into sports of all kinds.
>
> Obviously, I knew that people thought I was a girl, but that's not how I thought of myself. Even as young as age three or four, I felt like, "Something was very wrong. I don't know how or why my body is a girl's, but it needs to be fixed and someday I'll figure out how."

In elementary school, when I learned about puberty and what that would do to my body, I held onto a shred of hope that somehow puberty wouldn't happen to me. But at age eleven my breasts started to develop, and I did my best to hide them by binding them to flatten them. I used to wish for a medical condition or injury for which I would have to go to the hospital where my breasts would need to be removed by surgery. Until I was twelve, I didn't have any idea that there were other people who felt like I did, or that one day I'd be able to have my breasts surgically removed not because I was sick or injured, but rather for the real reason: they didn't feel like they should be part of my body. In the meantime, I cut my hair short and did what I could to present myself in a way that was consistent with the male person I knew myself to be.

My family was eventually supportive of my transition. When I was thirteen years old I started gender-affirming hormone therapy with testosterone. The primary feeling was relief; I was on my way to living in the right body and finally being able to express who I am.

I take my testosterone shot once a week. If it slips my mind and I'm a day late taking it, sometimes I'll notice a temporary decrease in my usually upbeat mood until I get to it. I had a mastectomy at age fifteen and a hysterectomy and oophorectomy [removal of the ovaries] a few years afterward. I'm happy with my life now. I'm engaged to my longtime girlfriend and have a career I'm passionate about. I'm still into exercising and staying fit. I have no regrets about my transition; it was 100 percent the right decision for me. The only thing I might change would be to have started it earlier, before I experienced the effects of estrogen so that maybe I could have gotten a few more inches out of my growth spurt. But I'll take what I've got.

If you met Alan, you wouldn't bat an eyelid (well, you might—he's very handsome). Alan keeps quiet about his history, and in ordinary social settings there's no danger of him being "outed." He's not a towering six-footer, but not unusually short, either. For Alan and Buck, T has transformed their lives.

BRICK HOUSES

Testosterone and related androgens (and also estrogen) direct the body's energy to be used to build up molecules and tissues, while other hormones, like cortisol and adrenaline, break down tissues and molecules to free up energy to fuel working muscles, among many other functions. Growing and remodeling the tissues of a boy into those of a man is no small feat of physiology. Accomplishing such a metamorphosis requires lots of energy, and it also requires tight coordination between the reproductive, nervous, endocrine, and metabolic systems. T is like the foreman of this massive construction/remodeling project, relying on a cadre of workers with different skills, and the right connections to ensure the smooth supply of a variety of materials. T recruits a team of hormones, including growth hormone, estrogen, insulin, and thyroid hormone, to help out, since they all have different areas of expertise. In general, these all play a role in deciding which tissues have priority at any point in life (e.g., growth hormone for childhood growth, T for muscle development in puberty, and progesterone to support uterine function in pregnancy). With T's oversight, the team can ensure that the right materials get deposited at the right times and in the right places for male reproduction.

Stopping the action of testosterone won't undo all its earlier work, even when supplemented with high doses of estrogen. Imagine building a brick house. Once they're up, the brick walls require little upkeep, but they are also difficult to renovate. The rest of the property, however, requires maintenance—regular painting inside and out, changing the AC filters, repairing the roof, watering the lawn. The construction projects directed by T are of both kinds: those that involve permanent changes requiring little upkeep, and those that need ongoing attention. Developing a boy's bone structure into that of a man, including growing the long bones, masculinizing facial bones like the jaw and brow ridge, or lengthening the vocal cords or, as specialists call them, "folds," are like building the brick structures, which are strong and stable but difficult to alter or tear down. But enhancing upper-body musculature, developing the reproductive system, and redistributing fat are like painting the siding and installing air-conditioning. Without maintenance and repair overseen by the T foreman, these other features will lose some of their previous function.

Testosterone's bricklike effects are the reason that physically transitioning in the male-to-female (MtF) direction is so much harder than the reverse (FtM). Many of the secondary sex characteristics that T produces in puberty, like broad shoulders, square jaw, and greater height, are obvious cues to the male sex and are difficult to eliminate or even significantly remodel or reduce. I've already discussed T's effects on bone in puberty (chapter 5), so let's take a close look at three other T-induced masculine features: a deep voice, protruding Adam's apple, and facial and body hair. MtF transitioners who have gone through a male puberty will generally find these features unwanted; FtM transitioners will generally find them desirable.

TONING DOWN

Remember the castrati from chapter 2, young boys with promising singing voices whose testicles were removed prior to puberty in order to prevent the development of an adult, masculine voice? Without the high T of male puberty, they retained the ability to hit the high notes.

My eleven-year-old son still sounds like a boy. When his voice deepens in a few years (usually around an octave), its sound will signal to me and everyone else that his boyhood is over. The quality of a person's voice, including breathiness, pitch, and intensity, provides others with a surprising amount of information—about sex, age, health, social status, even the stage of woman's menstrual cycle. A deep, strong voice is a potent signal of adult masculinity, sexually attractive, and a signal of dominance to other men.

Meet Kallisti. She transitioned in the opposite direction from Alan, from male to female. But she did this in her early thirties rather than at thirteen, and so endured the full suite of changes wrought by male puberty.

As a kid, I loved trying on my mom's clothes. I knew something about the way I was gendered was incorrect, but the people who told me I was a boy were in charge of things, important things like paying the bills etc., and I listened to them. But I knew it wasn't right. It wasn't really the clothes themselves. I think I understood even then they were just a costume. But as an actor, I know that a costume can act as a shortcut to expressing something deeper and richer. For me, that was the innate feeling that I was a girl. The clothes helped me to inhabit that feeling, express it. And it just felt . . . right. I felt somehow more myself.

Testosterone definitely had serious effects and influences on me when I hormonally transitioned as an adult.

Alan wished he could have experienced male puberty a little earlier to gain more height, and while Kallisti is a proud six-foot-four trans woman, her life would have been made easier had she not grown quite so tall or developed a masculine bone structure.

Puberty also gave Kallisti a deep voice. Although she can feminize it to some extent through voice therapy (as many trans women do), it still cues masculinity. I had to make an effort to override the masculine signal her voice sent over the phone, and it was easy to see why this could make life difficult for Kallisti.

The changing levels of several different hormones throughout our lives, such as estrogen, progesterone, growth and thyroid hormone, affect the quality of our voice by acting on the vocal anatomy, most prominently the larynx, otherwise known as the voice box. But none are as impactful as male levels of T in puberty, which are twenty to thirty times higher than in females of the same age. (The effects of sex hormones on the female voice during this same time are small, particularly compared to those during and after menopause when hormonal changes can cause the voice to become huskier.)

You can picture the larynx as a tubular structure at the top of your neck. Its base connects to your trachea (windpipe), another tube-like structure, that runs down from the larynx into the chest cavity before it branches into your lungs. The whole system allows air to

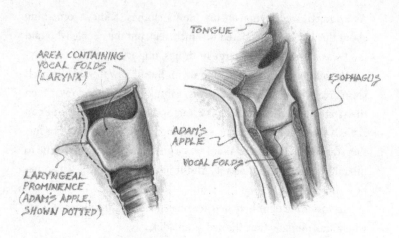

Vocal tract

pass between your lungs and through your nose and mouth; the larynx also acts as a valve that closes off the airway to protect it when
swallowing. All this tubery is obviously crucial to your survival, but
it also allows you to modulate the airflow so you can speak, shout,
or sing.

Inside your larynx are your vocal folds, a pair of short rubber-
band-like tissues that stretch across it. We can manipulate our vocal
folds to affect their rate of vibration and the sounds that can be produced. By relaxing and contracting the muscles that are attached to
the folds, we can alter the folds' shape, tension, and the amount of
space between them, sort of like a pair of lips stretching, closing, and
opening. The tissues of the larynx are rich in androgen receptors
during male puberty, and T interacts with its receptors to elongate
and bulk up those tissues. Among other effects, T causes the diameter
of the larynx to increase, creating a wider tube, and it also causes the
vocal folds to thicken and lengthen.

The length and thickness of the vocal folds are important determinants of the depth of your voice. If you play a string instrument the
principles will be familiar, but if you don't, take a rubber band, stretch

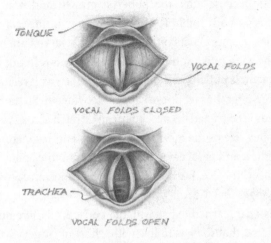

Cross section of larynx with vocal folds

it tight between your fingers, and give it a pluck. Then pluck it again when it has more slack and is thicker. You can also experiment with the length of the rubber band. Longer and thicker strings, bands, or cords vibrate more slowly when stimulated, producing a lower pitch, and shorter, thinner ones vibrate more quickly, producing a higher one.

Other T actions also help masculinize the voice, like building up the ligaments and muscles in the larynx and growing bones in the face to produce larger nasal and sinus cavities. T also acts to lower the position of the larynx in the neck during male puberty, giving the voice lower resonances (called "formant frequencies"). These changes all make it possible to project the voice more loudly.

T's actions on the vocal folds can't be undone by blocking T or taking estrogen later in life. Once thickened and elongated, the only way to bring them back to their previous state is vocal cord surgery. But the path to a manly-sounding voice is relatively smooth for FtM transgender people, no matter the age at which they start their hormonal transition. The voice will begin to drop within two to five months of starting male levels of T and will stabilize within a year. But it may never reach the same depth as that of a natal male.

This is because after the pubertal growth and remodeling of a natal female's body, high T, introduced later in life, can thicken the vocal folds, but its effects on the more stable larynx are more limited (similarly, wide hips cannot be narrowed). As a result of having gone through female puberty, the diameter of the larynx is relatively narrow, and high T after the fact doesn't appear to widen it. So the length of the vocal folds is restricted because they can't stretch across a wider larynx. The smaller size of the larynx, vocal folds, and resonating chambers, like the chest and nasal cavity, can all constrain the ability to achieve a deep, powerful voice. But overall, most trans men are satisfied with the vocal changes that T brings.

Kallisti has learned to live with her voice. And there's another masculine signal that she has to live with, short of surgery—the protruding Adam's apple. Again, that's all T's fault.

ADAM'S APPLES

Once you start paying attention to Adam's apples it's hard to stop (at least, it is for me). We women have them, but they just don't bulge out. You may wonder what the heck they are, and why they are bigger in men.

You may also wonder how the Adam's apple got its name. One popular origin story sounds as if it must be right: the name represents a bite of the forbidden fruit from the Garden of Eden that got stuck in Adam's throat. Unfortunately, that's a myth. Another theory comes from linguists and is that the Latin phrase that translates as "Adam's apple"—*pomum Adami*—is a mistranslation of a Hebrew phrase that means "bump on Man." But apparently that's not right, either. According to experts at Merriam-Webster, it all started with Arabic medical writers in medieval times, who called the throaty protuberance a "pomegranate." A complicated series of events then produced the English "Adam's apple." I'll just leave it at that.

Etymological speculation aside, we do know that the Adam's apple is the most visible evidence of one part of the larynx—the thyroid cartilage—that sits protectively over the vocal folds. The cartilage tissue is formed in two halves that join at the midline to form the Adam's apple—technically known as the "laryngeal prominence." The male larynx grows in response to high T during puberty, and the angle at which the two cartilages join is much sharper in men, due to the protrusion of the long (relative to women) vocal folds that the cartilage covers. In men, the two plates of cartilage join at an angle of about 90 degrees; in women, it's about 120 degrees. The larger larynx and pointy cartilage over the vocal folds both cause the Adam's apple to protrude more in people who have gone through a high-T puberty than those who have not. The Adam's apple may grow somewhat under the influence of T in FtM transitions that take place after puberty, but the effect is variable.

Finally, our last easily noticeable cue to gender—facial hair. This is something that many trans men want and T can deliver. Alan, who started T at age thirteen, has a neat, dark, short beard. Buck Angel, who waited until twenty-eight, has a beard that many men would be proud of.

GETTING HAIRY

Many of us think that we have too much hair on our body, or too little on our head. People with lots of testosterone are more likely to have both of those feelings. What you should definitely think is that your hair situation is abnormal among mammals. While the skin of most other mammals is protected by a thick, furry coat, in our natural state ours is largely exposed to the elements. Growing long hair on our heads is an equally bizarre feature of the human animal— even our close primate cousins have retained the typical mammalian hairstyle. The leading explanation for our hairless abnormalities is the "survival of the sweatiest" hypothesis. The idea goes that some-time after about seven million years ago, when our hominin ancestors left the African rain forest for the sunny savanna and adopted an upright gait, their need to efficiently dissipate heat increased. They swapped the fur coats for bare bods with extra sweat glands, allowing for more efficient cooling. Standing upright meant that less of one's body would be exposed to the midday sun, but the top of the head was now in a more vulnerable position and needed protection from the sun's rays. The solution? Lush locks on the top of the head, sparse hair everywhere else.

Before puberty, boys and girls have skin coated with soft, light "vellus" hair (also known as peach fuzz). As puberty approaches, in both sexes androgens are produced at a low level from the adrenal glands, which stimulate the development of pubic and armpit ("axillary") hair, in addition to the first hints of acne. But a few years later, ovaries and testes begin to secrete their products, and there's a fork in the pubertal road. The low-T feminine road, traveled by Buck Angel, spares many of those vellus follicles from further development, allowing them to continue producing the soft, slight hair. This is why women have much lighter hair on their legs and arms and, for the most part, no coarse hair on their faces.

The high-T masculine road, traveled by Kallisti, leads to what one prolific hair researcher calls the "most obvious biological marker of sexual maturity and biological maleness in our species." High T causes far more of those vellus follicles, particularly on the face and chest, to be converted into terminal follicles, which produce coarser, darker, "terminal" hair. Although men seem almost hairless relative to other mammals, compared to women, they are hairy beasts.

But it's not testosterone itself that is most responsible for manly hairiness. Recall Taman from chapter 4, the boy with 5-alpha-reductase deficiency. Although Taman had high T levels, because he had a mutation in the gene for the enzyme 5-alpha reductase, he could not convert the hormone into its potent derivative, dihydro-testosterone (DHT). Without DHT, not only was he unable to fully develop a penis during his early development, he was also unable to produce much facial hair as an adult.

Inside hair follicles everywhere but the scalp, the growth of terminal hair requires DHT. Inside the cells of hair follicles, testosterone is converted into DHT, which binds tightly to the androgen receptors there. In this way, DHT increases the transcription of the genes that promote the growth of terminal hair.

For trans men (natal females), increasing T from female to male levels will almost certainly result in a significant increase in body and facial hair, usually within the first year of hormone therapy. But some trans men just can't grow the robust beards that they might like (some non-trans or "cis" men have this problem, too). There are a few reasons for this. They may not have a sufficient number of hair follicles, or the follicles that they do have might not be sensitive enough to androgens, or perhaps they aren't making as much DHT as is necessary.

For trans women (natal males) who have gone through male puberty, T's action on hair cannot be easily reversed. Once T has converted those vellus follicles to terminal ones—enlarging them and increasing their androgen sensitivity—they can't ever fully return to their previous vellus state. Blocking testosterone and increasing estrogen (as part of feminizing hormone therapy) will slow the growth of new terminal hair, and may reduce the diameter of the hair that does grow, but that won't completely solve the problem. For trans women like Kallisti, these lasting effects of T can be distressing, bringing an urgent need to shave daily or even more often. For people who have the means, laser treatment or electrolysis is effective, and Kallisti uses such treatments to get relief from her daily struggle against those stubborn hair follicles.

Puberty leaves indelible marks on both males and females, which make physically transitioning after that more challenging. And, perhaps more important, there's the agony of puberty itself for people like Alan. When he was a kid, he dreaded the changes that would be brought about by a female estrogenic puberty. When he began to develop large breasts, it seemed like a nightmare. These external signals of femininity were a betrayal of how he felt on the inside.

Females enter puberty about a year earlier than males, and for many, their bodies have matured significantly even before they hit high school. Even though Alan began his T treatment relatively early, at age thirteen, it was too late to prevent the development of female secondary sex characteristics like breasts and wide hips.

Alan and Kallisti endured the puberty typical of their natal sex, developing unwanted traits in the process. Modern drugs could have prevented that.

BLOCKING PUBERTY

Along with the rise in visits to gender clinics, the demand for "puberty blockers" has skyrocketed. These are drugs that prevent the body from producing the sex hormones that direct the changes of puberty. Historically, they've been used to treat a serious medical condition called "precocious puberty," in which sex hormones rise well in advance of normal puberty. This leads to all the usual consequences—breasts, beards, bigger testicles, and elongated, thickened penises—even in children as young as three years old. Puberty blockers allow these children to continue normal childhood growth, and without premature sexual maturation.

Used for gender dysphoria, the idea is that blockers buy time for children facing the imminent development of traits that will take them further away from the body they want. When the blockers are stopped, treatment with cross-sex hormones can start if a physical transition is desired; otherwise, natural puberty will commence.

I spoke to Sasha, a twelve-year-old natal male, about Sasha's decision to delay puberty with blockers.

SASHA'S STORY

When I was little my parents always let me choose whatever I wanted to wear. I would wear pink clothes and stuff, I would always shop in the girls' section when I could. I mean, I was raised without these gender barriers set. Like when I wanted to wear something it wasn't bad clothes, it wasn't for one kind of person. I could wear what I wanted.

I went to a transgender sleep-away camp, and I've been going for a few years now. The first year I went there, I was still dressing really feminine, but still identifying as male and using male pronouns because I really just hadn't heard of anything else. But my friends that I made there, they were all going to the hospital, getting monitored for when they would go into puberty, and then they would be able to get blockers. And I really hadn't heard about it before.

So I just mentioned it to my mom in the car ride home from camp. Not even that I wanted it, just that it was this thing. And she said, "Oh, Sasha, I can make you an appointment at the hospital just to check it out." And then I started thinking about it and I realized that I don't want to go through male puberty at all.

Once I started getting older, like over the course of time in school, the boys started getting deeper voices. Adam's apples. I didn't want that, so I kind of realized, I don't want to look like that. If I went through male puberty I would get such a deep voice, I'd never be able to pass as anything but a man. And also puberty blockers are, like, temporary. I know that and going through male puberty is, like, permanent. So to me, it's more of a decision not to go through male puberty than to get blockers. So then I just wanted to get it blocked.

I dress and act, I guess, in very feminine ways. And I like doing that. And my pronouns are just "Sasha." Kind of like nonbinary. But super feminine. I'm not really one or the other, but kind of like neither. I feel mostly feminine, but I don't really like using the pronouns "she," "her."

When I'm in public and people say "she," I'm totally fine with it. But it's one of those things that if I know someone, they should just call me "Sasha." Like, if people say "he," which I used to be fine with because I didn't really care. But now when I think about it, like the way that I dress and act, I do care, and if I went through with a male puberty, my life probably wouldn't be very easy.

When I think about having a deep voice and getting hair on my body and having an Adam's apple, I don't like that. I don't like how it looks. I'm fine with what I was born with. But I don't want hair all over my body and stuff because then I couldn't really pass as female.

When I look in the mirror, I don't want to see, like, guy stuff.

What was your experience going to the gender clinic?

It was, like, a month after talking to my mom that I got the meeting. I was more leaning toward it, but not that much. So when they asked me, do you want a puberty blocker? I said, oh, maybe, maybe not. Not really sure. And I had, like, five other meetings and in the meetings I was, like, "probably," "most likely," "yeah, I really want to," "I definitely want it, I need to get it." Every meeting was like a little more once I thought about it. So we got a blood test to see how far along in puberty I was. So I got this surgery [to have the blocker implanted] shortly before I would have started going through actual puberty. It was really cool.

Going on estrogen after stopping blockers is probably what I'm leaning toward now, because I don't really, I didn't want to do that necessarily, but if the other choice is going through male puberty, then I have to, that's the only choice for me. I wouldn't mind having female traits, but the more important thing is that I don't have male everything. I wouldn't mind taking estrogen, it's just that the other options, of going through male puberty, outweigh everything else, I don't want that.

Sasha isn't sure what Sasha wants to be as a grown-up, but is leaning toward being a fashion designer or makeup artist.

PITUITARY HARASSMENT, BLOCKING T, AND PUBERTY

While some hormone blockers work by blocking hormone receptors, most puberty blockers work upstream: they stop the signal from the brain to produce the sex hormones in the first place. Normally, puberty begins when the brain starts sending signals to the gonads to produce sex hormones; and the sex hormones will continue to be produced for as long as these signals from the brain are received (most of adult life for men, until menopause for women). You might know the drill here, but just in case you'd like another refresher: the signals begin with the hormone GnRH (gonadotropin-releasing hormone) from the hypothalamus, which travels a tiny distance down to the pituitary, where it stimulates the release of luteinizing hormone (LH) and follicle-stimulating hormone (FSH), the gonadotropins. LH and FSH then travel to the gonads, stimulating them to produce and release estrogen and testosterone, among other sex hormones.

The drug Sasha is taking is the most widely used form of blocker, and ironically, blocks the release of sex hormones by activating the GnRH receptor in the pituitary. (The drug mimics the action of GnRH; it is called a "GnRH analog.") But how can that possibly help? Wouldn't the blocker just produce sex hormones from the gonads as usual?

Here's why the blocker works. Normally, GnRH hits the pituitary in pulses, and this pulsatility is necessary for the proper functioning of the whole system of communication between the brain and the gonads. The pituitary will only release LH and FSH into the blood when it gets its hits of GnRH from the hypothalamus at perfectly paced intervals, about one pulse every sixty to ninety minutes. It's as if the pituitary needs to be asked nicely, with breaks in between requests, to produce and ship out two of its main products. It will work diligently when proper protocol is observed, and each request is followed by adequate time to rest. But if GnRH breaks the rules and constantly bothers the pituitary, it just says "screw you" and stops responding altogether. LH and FSH won't be released at all, and the gonads will never get the message to produce their sex hormones. So blockers basically work by harassing the pituitary so much that it quits one of its main jobs, and in males T doesn't get made.

Blockers can be used for a maximum of about four years, starting in the early stages of puberty, around ten to twelve. Eventually, the decision will need to be made about whether to allow natal puberty to commence, or to give cross-sex hormones. These will initiate many of the pubertal changes that occur in the other sex, including the development of male- or female-typical secondary sex characteristics (like facial hair, deep voice, breasts, fat distribution, or muscle growth), but not including the development of the reproductive system.

SOME CONSEQUENCES OF HITTING PAUSE ON PUBERTY

Puberty blockers are drugs that interrupt the natural development of not only reproductive capacity but also any associated changes in the brain. We still have lots to learn about how sex hormones affect the brain in puberty, but it appears likely that this may be a second kind of period for brain organization, when estrogen and testosterone shape neural circuitry in potentially lasting ways.

Because the use of puberty blockers to treat gender-dysphoric children is relatively new, the research on long-term outcomes is almost nonexistent. We do know about some risks to keep in mind—social, physical, psychological, and reproductive. Children on puberty blockers are more likely to be shorter and less physically developed than many other children of the same age because only the latter group experiences the growth spurts of puberty. When children come off blockers later, normal physical growth will resume, but being developmentally out of sync with one's adolescent peers can be emotionally difficult. Puberty is also a crucial time for the accumulation of bone mineral density, and delaying puberty may cause irreversible losses in bone strength, although here again there isn't much evidence. But it's not just the changes in one's body that are relevant; they are typically coordinated with changes in how we feel about ourselves. And since adolescence is an especially important time for exploring one's feelings about one's gender, blocking puberty may also reduce opportunities to get more information as one matures physically, cognitively, and emotionally.

Coming off puberty blockers will allow a slightly altered natal puberty to begin. (It will start at a later age, and may not have exactly the same overall effects as a normal puberty.) Puberty blockers are then (largely) temporary and reversible interventions. The same cannot be said of cross-sex hormones, especially testosterone. When someone decides to hormonally transition, no matter their age, they are signing up for a lifetime of medical dependence on hormones (in addition to undergoing any gender surgeries they may choose, some of which are quite involved). The decision to experience a kind of cross-sex puberty carries more lifelong, serious consequences than the decision to take puberty blockers. But around 95 percent of gender-dysphoric children who use puberty blockers go on to hormonally transition.

Given the fact that almost all the kids who go on blockers end up transitioning with hormones, it makes sense to think carefully about future fertility in particular, prior to initiating blockers. If one moves directly from puberty blockers to hormonal transition, without giving the reproductive system time to mature, then the options for maintaining fertility appear to be quite limited. That is, if an FtM transgender person (a natal female) uses blockers to prevent natural ovarian development, and then undergoes a masculinizing puberty through the use of testosterone, he may never be capable of producing a viable egg. And similarly, if an MtF transgender person (a natal male) blocks testicular development prior to taking estrogen and experiencing a feminizing puberty, then her chances for producing viable sperm are low. But if one's natural puberty progresses long enough to allow the maturation of the reproductive system and the production of eggs or sperm, then even if one hormonally transitions later, it's possible for gametes to be harvested and cryogenically preserved (frozen). This procedure offers the potential for medically assisted reproduction in the future. And when a hormonal transition takes place after puberty and natural maturation of the gonads, if the gonads are left intact, fertility could return naturally when the individual discontinues cross-sex hormones.

I should emphasize that this book is not meant to provide any medical advice. Particularly where blockers are concerned, parents and caregivers should consult with qualified professionals, preferably getting a second or third opinion. But one thing is clear: to provide the best support possible to young people who are making these life-changing decisions, much more research is required.

TO T AND BACK

Most people are happy with their decision to hormonally transition. But there are some who are not. They may wish to transition back, and live again as members of their natal sex. At present there are no reliable data on the frequency of "detransitioning," but it is easy to find detransitioners who are willing to share their stories. I spoke to Stella (a detransitioned woman) about her three years on T, living as a man:

> I came out as trans right after I turned fifteen, then started my hormonal transition at sixteen.
> I was pretty lonely and depressed. I hated my body, my life, and had

trouble with my parents. I didn't have any friends, and I cried a lot. I wasn't gender dysphoric as a kid; it came on in my teenage years. I hated my female parts, wanted nothing to do with them. I didn't even want to acknowledge that I had boobs and a vagina, let alone look at those parts or let anyone else see or touch them. I had a sex drive, but I had no sexual interactions of any kind before I started on testosterone. I was trying to figure out my sexuality, but I knew I was attracted to women.

I got my testosterone through weekly injections that I would do myself. Immediately after my first shot, I felt like I had been injected with an instant antidepressant, and this response never changed throughout the three years that I was on it. Each injection felt like an instant mood boost, and it was easy to feel like I made the right decision. It was a physical reminder that I was in control. I felt so relieved that I would be able to escape the pain that (I thought) my body was causing me. And it worked really well for a while.

I had researched what to expect, and pretty much all that I expected happened. My voice dropped quickly, which was great. I grew hair where I wanted it, on my face, chest, and more on my legs (literally everywhere—stomach, chest, back, shoulders . . . and ass hair is a universal trans guy thing, just ask any trans guy). I could easily pass as a guy after less than six months. I started to love going to the gym and lifting, it was a big stress relief. In retrospect, I think that was one of the things that helped me realize I needed to detransition. I was taking care of my body and gaining muscle and looked super masculine, but I was still unhappy with my transition. No matter how well I was passing, eventually I knew that I could never get enough surgeries to feel completely comfortable as a trans man. It would never be enough for me, personally.

Lots of the problems I had before T were still there after T. Around the age of eighteen, I applied to college and didn't get in. I started asking myself what I wanted to do with my life, and I felt really confused. I had gained a bunch of weight, and then I started working out more and lost most of what I'd gained. I felt like, wow, I like my body! This was a big deal, because I hadn't had top surgery, and I'd hated my breasts before. I started to realize I didn't actually mind them, I only disliked them because they "shouldn't be on a guy." I slowly realized that my body was not the problem.

I hadn't realized that going back was an option, but I learned that I could just go off of T and let my body do what it does naturally. I also learned that the way my body operated on its own wasn't wrong, and I learned how to be comfortable with that. I could be who I wanted to be with the body I had, and I didn't need to change anything about it. So I went off T when I was nineteen, and back to the estrogen that my body produced.

Technically, I don't have to do anything physical to detransition—letting my ovaries produce their own hormones for the past year has led to enough changes in my appearance for me to mostly live normally as a woman again. But there are a few things that won't change back on their own.

I still have a masculine voice, and it sometimes prevents people from recognizing that I was born female, and I'm pretty upset by that now. Before I went on T, it had been the biggest change I was looking for from transition. My Adam's apple grew, so I plan on getting that reduced, because it's one of my biggest insecurities. I also have too much facial and body hair to deal with. I started the process of laser hair removal, which is annoying and expensive, but it's the only thing that will work for that thicker, darker hair. And my clitoris is still bigger than average, but that's totally fine.

Now I understand that I can never actually leave my body, despite changing how other people see me, so I wish I could have just learned to be happy with myself.

I'm now twenty, and a sophomore in college, and I love it. I'm much happier overall, though still struggling with my expectations of womanhood and being myself. I don't know what I'll major in, but I'll probably write about all this one day.

T AND SEXUALITY

Except for Sasha, who is still on puberty blockers, all of the people we've heard from in this chapter have talked about extreme changes to their libido, and this is one of the most prominent effects associated with T in transitions.

Let's hear from Alan (a trans man), Kallisti (a trans woman), and Stella (a detransitioned woman), about how their sexuality changed after they drastically altered their T levels.

Alan:

I had a healthy sexual appetite before and had always been attracted to women. This didn't change. That said, I did not want to be sexual while having an untransitioned, female body. It was very important to me that I transition and undergo masculinization to be seen as my real self before becoming sexual with a partner.

When I went on T, my libido definitely increased pretty quickly. Before starting T, I didn't understand the concept of an erection. But after starting T, I began having what I would call an analogous sensation that I'd describe as a sudden, intense feeling of pressure in the clitoris when I felt sexually aroused. Arousal also definitely happened more frequently than before starting T.

Kallisti:

As a teenager, and later, as an adult, before my transition, I was very into girls, but considered myself bisexual (now I identify as queer). I would not want to go through that first, testosterone-fueled puberty again, for any reason. It was a lot. It was years before I had any control over my sexual response. I'd be in math class, kind of zoning out, and find that I had an erection! What? It was very distracting. T seems like a heady drug, and the cliché is true—it really is like thinking with your dick.

But when I began my hormonal transition [blocking T and increasing estrogen], my patterns of sexual attraction changed, and I became much more interested in men, even though I was still attracted to women. And the experience of my sexuality changed dramatically. I wasn't consumed with thoughts of sex like I was before.

The loss of that intense libido wasn't something I minded; it was a relief, and I enjoyed sex much more. And not just because I was more comfortable in my body as a woman, even though that was part of it. Everything just worked better. Better orgasms—maybe less intense at the peak, but not as limited, not as confined to the genitals, and then just over. My whole body has become a sexual organ, sexually responsive, and the orgasms last longer and kind of affect my whole body. And I was pretty emotionally engaged and aware before, but now the emotional connection to my partner (my fiancée) plays a bigger role in my sexual response and enjoyment.

Stella:

As far as sex drive goes, I knew that it would go up, but it was like night and day. And after being on T for a few months, I went sexually crazy for guys, even though I still liked women. I spent more time thinking about who was looking at me, and wondering whether I liked them. If they appealed to me, it felt like a need that I needed to satisfy, like, right now. Being in a relationship was great, and I had really good orgasms on testosterone. They felt like a quick, concentrated release, as opposed to a slower, more full-body experience that I had before I transitioned.

My sex drive vanished for months when I went off of T but it's come back now, albeit in a very different way. Getting turned on is different. It was very obvious when I was turned on and the need to release was more urgent when I was on T. The release was also a lot more physical and gratifying after a single orgasm. I would become extremely sensitive when I had an orgasm on T. It almost feels incomplete now if I only have one orgasm. It's not a release, it feels more like the building of something bigger. I may be biased, but after experiencing both, I can certainly say I prefer the orgasms off T.

What Alan, Kallisti, and Stella told me confirms what's clear from the research on the experiences of transgender people who have hormonally transitioned: libido generally follows T. Transitioning from female to male using T grants entry into what can be a shocking new world of sexual desire that takes some time to adjust to. Male-to-female hormonal transitions often result in a lowering of libido. That doesn't mean a complete loss of sexual desire, but, with the addition of estrogen, sexual enjoyment may be softened and become more of a "whole body" experience.

We shouldn't jump to the conclusion that T causes these changes in sexuality via its effects on the brain. Transitioning brings a whirlwind of other physical changes, as well as psychological ones produced by being in control and on the road to the desired destination. Disentangling the neural effects of T on sexuality from all these other potential causes is a challenge. Still, changes in sexuality are remarkably consistent, and they precede visible changes in the body. It's not just a matter of enjoying sex more (which one would predict from a reduction in anxiety) but a qualitative change to a different way of being sexual. The T hypothesis fits with the data.

For the most part, these changes in sexuality are welcomed by the transitioner. Male-to-female transitioners report feeling relieved to experience a more feminine brand of sexuality, one that feels more consistent with how they identify; they don't seem to mind being freed from a preoccupation with sex. And for female-to-male transitioners, the same feeling is there, of experiencing a sexuality that feels more authentic. But for them, the manly version is often an eye-opener, as it was for Griffin Hansbury.

T AND EMOTION

There is far less research about emotional changes on and off T, but what there is suggests that the experiences of Alan, Kallisti, and Stella are fairly common.

Alan:

> As for anger, I wasn't at all angry before, and I'm not angry now. There's been no change there.
>
> Before T, I cried what I think is a normal amount. In retrospect,

I cried relatively easily. Now, I might have the emotion, but it doesn't come out in tears, even if I'd like to cry. I can have even more of that sad or moved emotion than would have led to tears before, but I still won't cry. It is usually years between episodes of crying. It can happen, but it takes a lot. The threshold is just much, much higher.

Kallisti:

I used to have anger issues. I was six feet, four inches with a full beard by my sophomore year in high school. I was a nerd, used to hang out in the A.V. club. I went through puberty early, and some short guys with Napoleon complexes used to try to prove themselves by taunting me, and it got to me, but I would never have physically lashed out at them. But I wanted to. I seriously fucked up some walls and doors when I would punch them in anger. I never got in a fight, never wanted to hurt anyone, but my anger seemed less rational and had more of a physical component to it before my transition. I once even lost a fight with a parking meter, and [messed up] my toe. I can't imagine doing something like that now. Overall my emotions seem more well balanced. When I transitioned and went off my natural T and on estrogen at thirty-three, I had crying fits and teenage temper tantrums, but that's all been balanced out.

Stella:

I could feel myself emotionally numb after T, but it felt normal after a while, just a part of me. I cried a total of three times during the whole three years I was on T, where I used to cry every day. I get happy and excited for things again. I didn't realize I missed that on T, because the dullness in my emotions was so normal. Anxiety, joy, depression, and excitement are all amplified in their own ways. The only emotion that seems duller now that I'm off T is anger, which used to be my most vivid emotion. I experience anger differently, more tied in with sadness than rage. I think most of my intense emotions before T were just due to still going through puberty, and it's balancing out for me now as an adult.

I asked Alan, Kallisti, and Stella specifically about crying, probably because I have my own issues with tearing up and I'm curious about it. They all experienced profound changes, in the predicted direction. In a recent opinion piece in the *New York Times*, Linden Crawford (a natal female) described the emotional result of being on T for a year. Crawford still had a desire to cry, but it was an impulse that "peters out before it reaches my tear ducts. There seems to be a thicker layer of insulation between my emotional core and my surface." Perhaps men and women feel similar emotions, but while women tend to let them bubble to the surface, men tend to let them stay put.

Women do cry far more than men on average, even though rates of crying among male and female infants don't differ. For the most part, as girls progress from childhood through adolescence, their crying doesn't change much. But the same cannot be said for boys, whose tears seem to dry up on the way to manhood. For me, the grass is greener on the other side. I wish I had a little more insulation.

Women not only cry more but they have higher rates of depression, and having lower T may have something to do with it. Stella reported getting an "instant mood boost" from her T injection. The evidence here is not straightforward, but some studies have shown that men with depression are more likely to have T levels on the lower end of normal (or even below that), and raising T under these circumstances may help to alleviate some of the symptoms of depression. But for men with healthy, normal levels of T who do not have a mood disorder, if there is a positive effect of T on mood, it is small. Perhaps Stella did experience a mood boost directly from her T shot. On the other hand, the placebo effect can also be powerful.

The changes in anger in T transitions are not consistent, which isn't surprising, since the frequency of feeling and expressing anger doesn't differ much between males and females. As we've seen, the sexes do differ in physical aggression, and T has a major role in explaining this. However, the fact that male-female T differences can help to explain differences in physical aggression doesn't mean that merely changing someone's T in adulthood will affect aggressive behavior. And most research on other human populations suggests that it won't. Aggression does not rise with increased T in natal men, even at very high doses, nor does it rise in natal men with testicular or neural disorders who go from a very low-T state (female levels) to normal male levels of T. And this is consistent with what we see in trans people. Admittedly Kallisti described a reduction in the desire to be physically aggressive, but maybe she just mellowed out with age, or maybe it's the happiness brought about by her transition. A number of personal or environmental factors could come into play. But in any case, there's no reason to think that T will turn a placid female-to-male transitioner into a hot-tempered Incredible Hulk.

LESSONS FROM TRANSGENDER EXPERIENCES

One important point to keep in mind is that trans men did not have high, male-typical exposure to testosterone in utero. And trans women did not have low, female-typical T levels in utero. Most transgender men and women went through the puberty typical of their natal sex. Furthermore, their social environment would most likely have differed from that of other people, sometimes from a very early age. For example, trans people may have behaved in ways that were less gender-typical than their peers and may have suffered social disapproval in response. All of this complicates understanding the effects of changing T levels as part of a gender transition; we can't simply assume that T would have the same kinds of effects in people who had very different early environments.

In this chapter, I've chosen to focus on the role of T in gender transitions. You may be wondering if natural variation in T levels has anything to do with being transgender in the first place. And given what we know about testosterone, that's a logical question. The answer is, we don't know. What little research we do have on the topic shows only weak and inconsistent evidence relating variation in natural T levels (or variation in the androgen receptor) to being transgender. The evidence that does link T to being transgender is the same evidence that links prenatal T levels to being a gender-atypical girl and then a lesbian, because childhood-onset gender dysphoria in natal females is strongly associated with sexual attraction to women. The pathways by which T affects gender-atypical behavior and (perhaps) sexual orientation in females might also affect the likelihood of being transgender. But for now, it's a mystery.

The transgender experience does have a lot to teach us about T. It's another piece of evidence to add to the pile, along with the evolutionary theory of sexual selection, endocrinological research on non-human animals and non-transgender human populations, and other evidence I have discussed throughout this book. Taking or suppressing T as part of a gender transition is the medically assisted counterpart of the natural case studies afforded by conditions like complete androgen insensitivity syndrome and congenital adrenal hyperplasia. In particular, trans people provide evidence that T is a crucial part of the explanation of male sexual behavior; it is not simply the product of upbringing or sexual standards prevailing in the culture. More generally, trans people's experiences with changing T levels confirm the conclusion of a vast and diverse amount of research that has been accumulating for more than a century: the power of T is truly awesome.

TIME FOR T

"MEN."

If you're a woman, there's a good chance that at some point you've said this to one of your friends. And you both understood exactly what it meant. Perhaps it was said out of frustration, during a conversation about the behavior of your husband, colleague, or a political leader. And, of course, men utter "Women" in a similar state of exasperation. We may even casually attribute vexing opposite-sex qualities to an endocrinological version of the "Mars and Venus" origins story: "Women are hormonal" and "Men have too much testosterone." My personal impression is that these somewhat playful "Men" sighs are a reaction to an objectifying sexuality, difficulties listening and expressing emotions, or ego insecurities that seem to motivate an unearned sense of confidence about, well, almost everything. This has the ring of sexism, so many of us let our feelings be known only in same-sex company.

Men watching too much football or talking over women is one thing, and it might very well have something to do with testosterone. But other much more damaging T-related behaviors have forced women (and some men) to shout loudly, "This is what is happening, and it must stop!"

Sexual assault is a complex phenomenon, but clearly sits at the crossroads of sex and physical aggression, two aspects of behavior in many mammals that are heavily influenced by testosterone. And the evidence that humans are included is strong. Some men may levy their greater social status and power over a woman to assault her; others may take advantage of a vulnerable woman who can't exercise any choice. Sometimes there's not much difference between those situations.

Given the gravity of sexual assault, we should take seriously all evidence-based hypotheses, and that means being open to the idea that testosterone is an important part of the explanation.

Solving problems requires understanding their causes. If we consistently downplay one set of potential causes (say, biological) in favor of another (say, social), then we have failed to do our best to get to the truth. And that means that we have also ignored opportunities to increase women's safety and equality between the sexes. We can do better.

One thing that women are doing better is speaking out about their experiences and shining a light on typically masculine brands of problematic behavior. Talking publicly about your sexual assault is not easy. Let's hear from one woman who did that, who reclaimed her story from the legal system, the popular press, and her assaulter. Her name is Chanel Miller.

CHANEL'S STORY

Chanel Miller is a writer and artist from Palo Alto, California. According to the police report describing Miller's 2015 sexual assault (when she was twenty-two), the perpetrator admitted he did not know his victim's name, nor would he be able to recognize her if he ever saw her again. In the widespread coverage of her assault, Miller was known in the press only as "Emily Doe." But that changed in advance of the publication of her 2019 memoir about the incident and its aftermath, *Know My Name*, when Miller revealed her true identity on an episode of *60 Minutes*.

I can't call Brock Turner her rapist, because in California, penetrating a vagina with one's finger—as Turner did—counts as sexual assault. Rape, in California, requires penetration by a penis.

On the night of the assault, two Swedish Stanford University graduate students, Carl-Fredrik Arndt and Peter Jonsson, were biking by a frat house on campus at about one in the morning. They noticed something amiss going on behind a dumpster. They could see two people lying on the ground, a man on top of a woman. He was "moving a lot" and she was moving "not even the slightest" bit, according to their reports. Jonsson shouted, "What the fuck are you doing? She's unconscious," at Turner, who fled. Jonsson gave chase, and Arndt stayed to check that Miller was still breathing. Jonsson tripped Turner and pinned him down, and the two men, joined by some passersby, restrained Turner until the police arrived and took over.

Turner was later convicted of sexual assault. At sentencing, Turner's father complained that his son shouldn't be sent to jail because prison is a "steep price to pay for 20 minutes of action out of his 20 plus years of life."

Miller read aloud her twelve-page victim impact statement at sentencing. Here's part of it, which also includes helpful instructions to would-be rapists and sexual assaulters:

> You took away my worth, my privacy, my energy, my time, my intimacy, my confidence, my own voice, until today.
>
> According to him, the only reason we were on the ground was because I fell down.
>
> Note; if a girl falls help her get back up. If she is too drunk to even walk and falls, do not mount her, hump her, take off her underwear, and insert your hand inside her vagina. . . . If she is wearing a cardigan over her dress don't take it off so that you can touch her breasts. Maybe she is cold, maybe that's why she wore the cardigan. If her bare ass and legs are rubbing the pinecones and needles, while the weight of you pushes into her, get off her.

The presiding judge, Aaron Persky, sentenced Turner to six months in county jail. His leniency was widely perceived as a miscarriage of justice, and in response Persky was recalled by California voters.

Most men are not Brock Turners. Many men are not as heroic as Carl-Fredrik Arndt and Peter Jonsson—trying to physically restrain another man can result in serious injury. As we've seen before, sex differences are largest at the extremes, and sexual assault and taking risks to help others are no exception.

And, yes, the stereotype of men as more likely to perform dangerous acts of heroism is accurate. Women deserve credit for other kinds of physical acts they perform routinely that can also be considered heroic—like bringing a new life into the world by pushing something the size of a watermelon through something the size of an avocado—but physical bravery on behalf of others is one area in which men excel.

Since 1904, in the United States and Canada, about ten thousand people have been awarded the Carnegie Hero Medal, which goes to a civilian "who voluntarily risks his or her own life, knowingly, to an extraordinary degree while saving or attempting to save the life of another person." About 10 percent have gone to women. Recipients have saved people from drowning, from house fires, from animal attacks, and so on. For example, one of the fifteen winners in 2020 was fifty-two-year-old Winston S. Douglas, a bus driver from Atlanta, Georgia. While on his daily route, he witnessed a man stabbing a middle-aged women as she tried to cross the street in front of him. Douglas stopped his bus, jumped out, and attacked the assailant, whose knife broke off as he tried to use it to stab Douglas. Thanks to Douglas's quick action, the victim survived. The sole female winner, Yolanda Robinson Isom, died after reentering her burning house to save three of her sons.

You might argue for an alternative explanation for the gender imbalance: obviously, sexism!

These days, luckily, people are more attuned than ever to gender imbalances. Granted, the people on the Carnegie review panel—who evaluate the cases and decide on the winners—are mostly men, but one-third are women. And there's not much they can do about the sex of the people who are nominated, who are overwhelmingly male.

To repeat, we are talking about extreme cases, where sex differences in behavior are relatively large. Women also risk their lives to help others (not to mention participating in dangerous sports like mixed martial arts and auto racing). But the fact is that men from diverse cultures seek out thrilling, intense, novel, and adventurous activities and take more physical risks than women. Testosterone is likely implicated.

In her statement, Chanel Miller made sure to recognize her heroes: "Most importantly, thank you to the two men who saved me, who I have yet to meet. I sleep with two bicycles that I drew taped above my bed to remind myself there are heroes in this story."

Two years after Chanel Miller's ordeal, the journalist Ronan Farrow published an article in the *New Yorker* documenting allegations of sexual assault by the movie producer Harvey Weinstein—an article that started a cascade of events that would see many men accused of sexual assault and worse.

#METOO

In 2017 Weinstein's star power was unrivaled; as Farrow noted, Weinstein's movies, from *Pulp Fiction* to *Shakespeare in Love*, "have earned more than three hundred Oscar nominations, and, at the annual awards ceremonies, he has been thanked more than almost anyone else in movie history, ranking just after Steven Spielberg and right before God." The *New Yorker* went on to reveal that thirteen women, including the actresses Mira Sorvino, Rosanna Arquette, and Asia Argento, had accused Weinstein of sexual assault and harassment. It didn't take long for the numbers of Weinstein's accusers to swell. The #MeToo movement had begun.

By the time Weinstein was convicted of rape in New York in 2020, and sentenced to twenty-three years in prison, #MeToo had snared hundreds of other powerful men (predictably, along with only a handful of women). Among them was the comedian Louis C.K., who had a habit of masturbating in front of women, sometimes during meetings in his office. Reflecting later, he astutely observed that "what I learned later in life, too late, is that when you have power over another person, asking them to look at your dick isn't a question."

One immediately striking thing about the #MeToo revelations and Miller's assault is how unsurprising they are. Powerful, older—and often quite unprepossessing—men are exploiting their positions, through whatever means necessary, to have sexual encounters with a series of younger and attractive women. A young frat boy took what he wanted from a woman who was unable to say no.

What would be surprising is a stream of similar allegations about women. True, powerful women in the media, politics, and industry are greatly outnumbered by powerful men, but the women seem to be just as capable of using their power to get what they want. And it's not usually to have a constant supply of boy toys, to coerce men into sex, or to masturbate in front of them. And when is the last time you heard of a woman caught in the act of fondling the genitals of a man she didn't know who was passed out on the ground?

Most men I know are not Weinsteins or Brock Turners. Weinstein is an outlier, but he is no freak of nature. He is a man whose combination of power, sense of entitlement, personality, libido, and opportunity gave rise to a perfect predatory storm. Testosterone tends to promote high libido and the acquisition of mates, and if a man's power, a culture's failures, or a victim's powerlessness can do the trick, that will be the road taken by some. But we can also put roadblocks in the way.

#MeToo is a movement that has made real progress, and hopefully that will continue. As I have emphasized throughout this book, to bring about changes in male behavior it is not necessary to depress testosterone. Changes in attitudes and culture can do that all by themselves. As Steven Pinker reports in his book *Enlightenment Now*, in the United States the "rates of rape and violence against wives and girlfriends have been sinking for decades and are now at a quarter or less of their peaks in the past." Declining male T levels are not the explanation. Even though, as far as we can tell, men's heightened desire for sex has not changed, what has changed (in some places) is the sense of entitlement of some powerful men.

The social practices of reward and punishment, praise and blame, affect the expression of human behavior. We still have a long way to go, but understanding T can only help to identify structural reforms that will promote positive change.

T DIVIDES

No two animals are exactly the same, not even identical twins. This is ensured by variation in the environment, by variation in how genes are expressed, and by random effects on developmental trajectories. But only a few of us have an identical twin. The rest of us will come into the world with a unique complement of DNA, which makes for differences among people. Despite these genetic differences, all humans are very much alike. Still, looking at humanity as a whole, humans come in two visibly distinct forms: male and female. This is due to a small genetic difference: the SRY gene, found on the Y chromosome. Without that gene, the body almost always develops as female.

That humans come in these two forms has no implications for the value of a person's life. For instance, there are no implications for people whose primary or secondary sex characteristics don't align as usual with their sex chromosomes. By the same token, there are no implications for many people who transcend social expectations associated with their sex.

As we've seen, the SRY gene cleaves the sexes apart because its activity results in testes, and high testosterone.

T is broadcast through the blood, mostly from the testicles (in males), and received by androgen receptors in cells all over the body, including the brain. Once the T-key is in the androgen-receptor lock, the T-lock combo migrates to the cell nucleus, where it increases the rate at which certain gene recipes are read and new proteins are baked. (Which gene recipes are affected depends on which pages of the gene-recipe book are accessible, which in turn depends on the type of cell, among other factors.)

The primary evolutionary function of testosterone is to coordinate a male's body and behavior in the service of reproduction. To that end, early T exposure in many male mammals (including humans) masculinizes the brain, influencing the development of neural circuitry that is activated later in puberty.

Testosterone changes everything. It changes the way genes are expressed on every chromosome—proteins from thousands of genes are produced in systematically different patterns and quantities in men compared to women. These proteins go on to affect the body and the brain, first in utero, again shortly after birth, and then there's another explosion of changes in puberty. T affects body and behavior, which in turn affect one's social environment, which in turn affect body and behavior, and so on and so on, until death. T divides.

T DOMINATES

Some T skeptics say that calling testosterone the "male sex hormone" is wrong because women have T, too. But since it is responsible for making and maintaining male bodies (with a far smaller role in female ones), the name appropriately conveys its actions.

As I have explained throughout this book, the effects of T are deep and wide-ranging. Boys' preference for rough-and-tumble play and men's motivation to compete with other men, their greater libido and preference for sexual novelty, and their athletic advantage over women are all testosterone derived. T is also involved to some extent in sexual orientation, although here its exact role is at present unknown. The strongest evidence links prenatal T exposure to same-sex attraction in women. Prenatal T is also related to other sex differences, such as in career preferences.

T levels in healthy men and women do not come close to overlapping: men's are 10–20 times those of women's. In puberty, the gap in T level is even wider—pubertal boys have about thirty times as much T as girls. T levels change throughout a male's lifetime, starting when he is a grape-size fetus with minuscule undescended testicles. T levels are affected not just by what goes on inside the body but also by what goes on outside it.

The social environment can bring about changes in T levels within minutes. Elevated T after winning a contest, or reduced T after losing, may allow a contestant to respond adaptively, provided he has the appropriate constitution and is in the right circumstances. Winners should capitalize on their success and be primed to face a challenge that may require physical action, while losers should be more cautious, next time picking on someone their own size or simply retreating in the face of a threat. The effects of these short-term changes depend on personality, not just T. They also depend on the density and action of androgen receptors, which in turn depend on genetics. T does not have the same effect in everyone—genes, status, personality, health, and the present social situation all matter, too.

T affects the social environment, through its direct effects on the brain and behavior. But it can also affect the social environment indirectly, through its effects on the body. Simply possessing certain physical traits, like a deep voice, large body, or big muscles, profoundly affects social interactions. Transgender people, crossing from one side of the T tracks to the other, know this only too well.

Fatherhood, in cases where dad is involved with the little ones, produces a longer-term change in T. Keeping T high has its costs, which is why T tends to drop when a man (or sparrow, stag, or spiny lizard) needs to direct his energy away from competing with other males and toward maximizing his own health or that of his family.

CONVICTING T

If a detective is trying to solve a crime, she would be unwise to rely on a single source of evidence: an eyewitness, DNA left at the crime scene, or a suspect's confession. Eyewitnesses can be mistaken, DNA samples can be contaminated, confessions can be coerced.

Multiple independent sources of evidence can combine to strongly support a hypothesis, whether it's about the cause of a rattle in your car, why your soufflé has collapsed, or why someone blocked you on Twitter. It's just like that in science. And I've tried to indicate throughout this book how multiple independent sources of evidence converge to justify these conclusions about T.

First, there's evolutionary theory—about as solid as anything in science. We have been sculpted by the processes of evolution just like any other species. We have compelling evidence that human sexual dimorphism is largely the result of sexual selection: an evolutionary process that favors traits that improve the ability to acquire mates, either in quantity or quality. Men's and women's reproductive interests are similar but not identical. Given the constraints of mammalian biology, women are the more heavily investing sex—they are required to spend significant amounts of time and energy to produce a child, while men may accomplish this with a far smaller investment.

Like women, men benefit from competing with others of their sex for the highest-quality mates, but there are some important differences. For men, high social status is more reproductively advantageous than it is for women. Direct forms of competition, including physical aggression, help men to reach higher on the totem pole. Men also have a greater sex drive and preference for sexual novelty. All of this implicates T.

Second, there's behavioral endocrinology, which has come a long way since its nineteenth-century origins, when Arnold Berthold transplanted testicles into the abdomens of castrated chickens. We have discovered the chemical structure of hormones, how they act on receptors and affect gene transcription, and so much more. We understand in exquisite biochemical detail the mechanisms by which these simple molecules work their physiological magic.

Third, there are natural conditions, like complete androgen insensitivity syndrome and congenital adrenal hyperplasia, that give researchers an opportunity to observe what happens when Mother Nature drastically alters exposure to androgens. The evidence points to T as a potent masculinizing force.

Fourth, there are artificial experiments, like the production of eunuchs in ancient China, and castrati in sixteenth-century Italy. Much more recently, there is the medical suppression or elevation of T as a part of a gender transition.

All that said, there *are* alternative hypotheses to the ones I have advanced in this book. Notably, there's the hypothesis that sex differences in aggression are due mostly to socialization rather than evolution (with T as an accomplice). I've suggested throughout this book that socialization does not fare well against its evolutionary T-based rival. Here, I think the psychologist Steve Stewart-Williams has summed the situation up nicely:

> How would socialization theories explain the violence gap that opens up between the sexes at puberty? Is there a sudden surge in gender socialization—a surge which, for some unknown reason, happens at exactly the same stage of life in every culture and in many dimorphic species? Is it just a coincidence that this alleged surge in socialization comes at the same time as the massive surge in circulating testosterone that accompanies puberty in males?

T SKEPTICISM

The popular press is full of attempts to bring King T down, to show that he is too big for his boots, or more generally to dismiss biological accounts of psychological and behavioral differences between the sexes. For example, in 2020, the *New Yorker* published an interview with the author and journalist Peggy Orenstein, about her latest book, *Boys & Sex: Young Men on Hookups, Love, Porn, Consent, and Navigating the New Masculinity*. In the article, titled "Can Masculinity Be Redeemed?," Orenstein was asked whether she was "making an argument about biology" when she talked about the things young men needed and valued, like "athleticism, dominance, and aggression" and "wealth and sexual conquest." She laughed and said:

> No. What we know is that nurture becomes nature. You see that all the time. The sorts of things that we learn as small people or from the media that we absorb or from the messages we get from family shape whatever that clay is of how we are born.... Kids are bombarded with messages in the media about male sexual entitlement and female sexual subjectification. We have long recognized how those messages can undermine girls' self-esteem and reduce girls to

their bodies and affect their cognition, all these different things. But boys are cooking in that same stew, and I would argue in some ways the temperature is higher.

Those who think that T is an important factor often find their positions caricatured. Here's an example from a *Guardian* review of Cordelia Fine's book *Testosterone Rex*. The journalist reviewer writes that because "testosterone makes men tall, hairy and deep-voiced," it is easy to see how knowledge of T's physical effects could lead us to imagine that it also produces other masculine-seeming traits, such as "leadership, violence and horniness."

However, the reviewer continues, Fine shows this idea to be a "fiction." The violent horny male leader, it turns out, is a product of socialization, not of testosterone's actions in the brain. Instead, those people who are born male, who show the effects of testosterone on their bodies, are treated differently from day one, and socialized into aggression, casual sex, and a sense of entitlement:

> Being male or female isn't enough to make you into your society's version of a man or a woman. . . . But as soon as your maleness or femaleness is recognised, other people start to treat you in ways that form you into a man or a woman, with the support of toys, books, role models, and a million other subtle nudges.

Notice how the journalist characterizes the view she thinks is wrong: simply being born with the body of a male or female is "enough to make you into your society's version of a man or a woman." She's right! Whatever a society's version of gender is, obviously all males don't fit the masculine paradigm, and all females don't fit the feminine one. Who on earth thinks that? No serious scientist ever said genes, hormones, or genitalia alone explain differences in any complex behavior. As we saw with aggression, culture is undeniably an important factor, and the same goes for sex. (Fine herself, I should emphasize, is a much more precise and careful writer than many of the reviewers of her book.)

Why this almost visceral hostility to those of us who talk about the powerful and important effects of T on male traits and behavior? It appears to have its source in three main worries. First, people think that the T view suggests that testosterone is destiny. Second, they think it suggests that male behavior is natural, and thus good or acceptable. And, third, they think it suggests that men aren't to blame—their T gets them off the hook.

T FATALISM

In the *Guardian* review of *Testosterone Rex*, the reviewer seems to think that if T divides the sexes, then there's nothing that can be done to curb men's excesses: if "hormones make the man or woman, and we are what we secrete, then efforts to end male dominance would be futile at best and possibly downright harmful."

The first point to emphasize is that even if fatalism is the outcome, then although this is a good reason for getting depressed, it is not a reason for rejecting the conclusions of this book! Shortly after I graduated from college, doctors diagnosed my father with incurable pancreatic cancer, which gave me a good reason to bawl my eyes out, which I did. But my response was not to argue that the doctors were wrong. Sometimes the news is bad.

But in this case, the news really isn't bad, and we can change the situation. Among healthy humans, with rare exceptions, there is not one gene or even one hormone that unilaterally causes anyone to act in any particular way or that sets anyone up for a specific kind of future. For example, depression runs in one side of my family, and I've struggled with it myself. I may have genes that predispose me to depression, but my behavior has a big impact on whether I feel the symptoms. I can exercise, prioritize family, and engage in fulfilling work, all of which help enormously (this isn't true for everyone suffering from depression). People who have a genetic predisposition to type 2 diabetes can often make changes to their habits and environment so that they don't suffer from the effects of the disease. And they have to stay on top of it.

Learning about how depression works, and that it has a genetic underpinning, has empowered me to make changes that help. I know that I have a propensity for it, and it's unlikely that any environmental change will forever cure it. If my attention lets up, and I don't pay attention to my symptoms and ensure that I'm doing what I can to keep them at bay, they will come back.

It's important to realize that changing our environment can actually alter what goes on inside our bodies—the way our genes are expressed, the levels and actions of our hormones. If I exercise every day I will change my dopamine levels; and if I eat less sugar my insulin won't spike up; if I'm a man and I get into the boxing ring for a fight, my T levels may rise, and so on. Learning about how the intertwined forces of genes and environment interact can help us understand the causes of behavior and make beneficial psychological and social change easier, not harder.

T: NATURAL AND GOOD

There's another explanation for the resistance to looking at evolution, genes, and hormones as major contributors to undesirable male behavior. If that's where the roots of violence and sexual assault lie, doesn't this mean that the worst of male behavior is natural and thus acceptable? Similarly, if men's greater striving for social status is the result of evolution and T, doesn't this justify existing inequalities between the sexes?

I hope you're thinking: *No, it doesn't!* But perhaps you shouldn't congratulate yourself on your critical reasoning skills. When it comes to traits or behaviors that you want to validate, this sort of nature-is-good idea might well be more appealing. We are more likely to embrace (and even search for) biological explanations for behavior when we approve of it. This is well illustrated by the lyrics of one of the top-selling pop songs of all time, Lady Gaga's "Born This Way." Gaga tells us that whether we are lesbian, transgender, gay, bisexual, or straight, "God makes no mistakes," and we are beautiful, because we are "born this way." I don't suppose "brought up this way" would have the same ring to it.

The song has become an anthem for LGBTQ people in particular. In an interview with National Public Radio, a young gay man described how the song made him feel, as someone who hasn't felt accepted for who he is: "All of a sudden, the idea that you were born this way and can't change who you are isn't just something that you feel: It's something the entire world is being forced to understand."

If homosexuality is produced by a "gay gene," or is found in many nonhuman animals, this line of thinking goes, then it's not a choice: homosexuality is a product of nature, and so should be accepted. You can find a similar idea these days about transgender people: if trans women (say) have some hormonal or genetic condition that causes them to have female brains in male bodies, then being transgender is something that's a product of nature, and so should likewise be accepted.

This is sometimes called the "naturalistic fallacy." As Steven Pinker says in *The Blank Slate*, the naturalistic fallacy is "the belief that whatever happens in nature is good." It's not called a "fallacy" for nothing. "Natural" diseases like malaria are not in any way good. Nature is full of wonderful things, but it's also full of truly awful things. Whether you're born to be gay, choose to be gay, or are gay because of your upbringing has no bearing on whether it's good to be gay (this applies to any aspect of who you are). Exactly the same point goes for men's aggression or heroism. Pinker puts it neatly:

> As soon as we recognize that there is nothing morally commendable about the products of evolution, we can describe human psychology honestly, without the fear that identifying a "natural" trait is the same as condoning it. As Katharine Hepburn says to Humphrey Bogart in *The African Queen*, "Nature, Mr. Allnut, is what we are put in this world to rise above."

For some reason endocrinologists seem to be generally immune to the naturalistic fallacy. Perhaps they are temperamentally an unsentimental and hardheaded bunch. Sheri Berenbaum is a professor of psychology and pediatrics at Pennsylvania State University, who has devoted her career to researching the behavioral effects of androgens on the developing brain. She forthrightly says:

> Rather than rejecting biological differences, we must seek to reveal the nonsense in the arguments that brain and behavioral sex differences justify discrimination, segregation, and differential treatment of the sexes.

I could not agree more.

T MADE ME DO IT

Finally, one more explanation for the resistance to biological explanations of sex differences. Some may worry that if people believe that T is responsible for reprehensible male behavior, then it provides men with a free get-out-of-jail card. And there is a basis for such a concern: people are more forgiving of people's behavior when they think the explanation for it lies in genes or other aspects of one's internal state rather than the environment. But not always.

The "roid rage" defense is sometimes tried in courts. In 1988 Horace Williams, a bodybuilder who took large doses of synthetic androgens to increase muscle mass, was charged with the brutal murder of a hitchhiker. His attorney pleaded "steroid-induced insanity." But it didn't work: the jury apparently thought that Williams could have chosen otherwise. They found him guilty of first-degree murder, and he was sentenced to forty years. His attorney expressed disappointment at the verdict: "There's no question in my mind the steroids are what caused (Williams's) violent behavior. Horace is not a violent guy. They made him insane."

The question of whether our genes, hormones, neurotransmitters, or what we had for breakfast excuse our behavior raises rather large philosophical issues about free will and responsibility that are not the topic of this book—nor are they in my area of expertise. But I can ignore these with a clear conscience because this kind of argument isn't really about T. Suppose Daemon Fairless's fight with the drunk man on the train that I described in chapter 7 only happened because T rose in Fairless's blood in response to a perceived challenge to his status. Does that give him an excuse? Well, there has to be *some* kind of biochemical explanation of his actions. If Fairless throws a punch, that's not magic—something has to happen in his brain that causes the impulse or decision. So suppose it isn't T after all: it's brain factor X. The fight, then, only happened because of X. Does *that* give Fairless an excuse?

That's just to say that this worry about T excusing male excesses actually has nothing to do with T. It's a general worry about whether biochemical causes of our actions excuse them. Insofar as this is a problem, it's a problem for everyone.

BACK TO THE SEMINAR ROOM

I'd like to revisit my experience in the seminar that I described in the first chapter. I felt angry and upset about a paper by Randy Thornhill in which "rape" in the scorpion fly was used as a jumping-off point to hypothesize about the evolution of rape in humans. When I was asked for my feedback, I told the rest of the group that I thought the author, whom I had never met, was an asshole.

Every time I think about that moment, or describe it to my students (which I do every time I teach about similarly sensitive topics), I have lots of emotions. Some things that should be blindingly obvious in the moment are only so in retrospect, and that's the somewhat embarrassing situation I've found myself in, here at the last chapter. Rape is a difficult subject for almost any woman because most of us fear it and because so many of us have been through it. I'm one of them.

I took that graduate seminar when I was thirty-five, many years after being traumatized by sexual assault. It's only through writing about men and testosterone that I have come to appreciate that my driving desire to learn about testosterone and how it works might have something to do with my own difficult experiences with men. But it hasn't been all bad: while some men have wounded me, far more have supported, mentored, and encouraged me.

Why did I resist an explanation for human rape that situated the cause in our evolutionary past? Because that explanation seemed to make rape natural and excusable, and that galled me. But equipped with an understanding of evolutionary biology and some basic logic, I came to appreciate that such a conclusion is not warranted. That doesn't mean the hypothesis that Thornhill offered is correct, but it does mean that I'm capable of evaluating the evidence on its merits, without letting my emotions get in the way. I've found this to be immensely empowering.

During that seminar, I wasn't coming unglued, but I was tearing up a little (which wasn't a surprise to anyone at that point) and struggling to find the right words. Perhaps the whole episode could have been avoided had I been warned about the topic of the paper and the discussion before I encountered any of it. That way, I could have mentally prepared myself for potentially disturbing content, or even skipped class entirely. And when the professor saw me in distress, he could have furrowed his brow with concern, offered me a tissue, and focused the discussion on something else. But he didn't do any of that. Maybe his being British had something to do with it, but he clearly thought I should keep calm and carry on. I was supposed to dispassionately consider the evidence and the argument. Don't get me wrong, I'm still pissed off. But I have better places to direct my anger than against the well-intentioned author of a scientific study.

The primary emotion I feel when I look back on that day now is gratitude. Like many students who have impostor syndrome, deep down I didn't feel that I belonged at Harvard or in that classroom. But here was a professor I admired and respected, patiently expecting me to do what scientists do. This was one of the most valuable lessons I learned in all of my training at Harvard, and I can't imagine a more effective way to have learned it.

HATE ALL MEN?

What's the appropriate response to the fact that men do the majority of raping and assaulting, not to mention hoarding the world's power? One option was suggested by Suzanna Danuta Walters, professor of sociology and director of the Women's, Gender, and Sexuality Studies Program at Northeastern University, in an opinion piece in the *Washington Post* in 2018:

> Pretty much everywhere in the world, this is true: Women experience sexual violence, and the threat of that violence permeates our choices big and small. In addition, male violence is not restricted to intimate-partner attacks or sexual assault but plagues us in the form of terrorism and mass gun violence. Women are underrepresented in higher-wage jobs, local and federal government, business, educational leadership, etc.

The headline was: "Why Can't We Hate Men?" The answer: we can! Hating men is actually the appropriate response to their litany of sins. As Walters bluntly put it: "We have every right to hate you. You have done us wrong."

I'm not happy that the *Washington Post* deemed this vilification of half the human population worthy of publication, but I am glad that I live in a country where people have the right to freely express such views. Not surprisingly, Walters's article generated plenty of controversy and blowback, including threats of gruesome violence directed toward her from men only—also not surprisingly. But she had her defenders.

Lots of books these days are aimed at inspiring young girls to shoot for the stars, to be fierce, tough, smart, and strong, and describe the accomplishments of women who embody those characteristics. And encouraging girls to aim high is a good thing. At the same time, men are blamed for simply existing, for their inherent toxicity. It's true that men are different—being scholarly, I should add "on average." But let's also not forget about the virtues that nature seems to have preferentially handed out to men. They might sometimes feel the need to confidently explain the obvious, but they also put their lives on the line for others and are massively overrepresented in the most dangerous occupations. The Ugandan men with whom I trekked through the jungle for eight months protected and educated me. Without them, this book would not exist.

So what *is* the right response to the male propensity for sexual assault and other problematic male behaviors? Let's not forget that men themselves don't get off easy, since they are the primary victims of male violence. Women are not the only victims here. The answer is not to try to take down the science that helps to explain the observations. It is not to hate men or the testosterone that runs in their blood. We already have some ideas about what works—tremendous progress has occurred in my lifetime, and even in the last few years. Scientists can and should do a much better job of educating journalists and the public about the huge body of findings about the biological roots of sex differences. This research is not only fascinating but it can be life changing. Learning about science from a talented and caring teacher helped to instill in me a lifelong love of using the tools of science to understand myself, and the wondrous, and the worrisome, aspects of human behavior.

Science, of course, is not the only way to understand ourselves. Books, music, the visual arts, poetry, travel, other people, and ideas that take us out of our comfort zones—these are all ways for us to learn about humanity. But science in particular—including a basic knowledge of statistics, hypothesis testing, biology, logical reasoning—can equip us with the tools we need to intelligently process the enormous amount of information with which we are confronted every day. When good causes are mixed up with bad science, or when propaganda and conspiracy theories have more sway than good data, something has gone badly wrong.

BACK TO THE FOREST

Remember Imoso, from the first chapter? He was the dominant male who beat up my favorite female chimp, Outamba. Imoso held most of the social power and had a higher than average amount of testosterone. He achieved his high status partly through his ability to form loyal coalitions but also because he was quick to anger and physically aggressive toward those who did not properly display their subordinate status. And he was particularly brutal toward the adult females. For him, this was a solid evolutionary strategy.

The genes that Outamba and Imoso carry aren't so different from our own, and in many ways we can see echoes of our behavior in the behavior of chimps. One of the frightening things about Imoso's attack on Outamba was that it was all too human. Chimps are not aliens to us, in the way fish or insects are.

Since that time, I've learned and thought a lot about the ways in which humans are unique, and different from chimps. Here, for me, are two headliners.

First, sex differences in humans are relatively small compared to chimps and many other species. One big reason is that human children tend to do better with two parents, and male-male competition isn't as intense in humans as it is in chimps, or other species like red deer or lizards. Second, we humans have evolved big brains and the capacity to reflect on our choices. Relatedly, we have constructed fabulously sophisticated societies, which serve as a repository and generator of knowledge, to be transmitted and extended through the generations. Unlike chimps, we can learn about our evolutionary origins and the resulting biochemical mechanisms in our bodies. That gives us power over ourselves that no chimp comes close to having.

"MEN," REVISITED

Let's get rid of the tired idea that the sexes must be born with basically the same brains in order to have equal rights. (That usually seems to mean that women should be more like men, and perhaps sometimes motivates the search for more significant T action in women.) Men and women, and boys and girls, are different, and that is in important ways traceable to differential exposure to androgens, starting in the womb and continuing in later life. But of course individuals vary. My son Griffin never played with trucks, and I used to dismember my neighbor's Barbie dolls (Kirsten, I'm sorry!). Still, there are consistent patterns.

When Griffin was little, I was curious about why he didn't want to play with trucks and blocks like lots of other boys. He could play with "girls' toys" or "boys' toys"; it was all the same to my husband

and me. Actually, for his first few years in school here in Cambridge, Massachusetts, he was positively rewarded for playing in gender-atypical ways. (And dressing that way, too, sometimes—if you know progressive Cambridge, that won't be a surprise.) But he still wanted to pounce on his friends and draw comic after comic about bad guys blowing things up with the good guys coming to the rescue. Research on play preferences helped me to learn what was typical but also informed the way we responded to his choices. We tried to be as open and supportive as possible, knowing that even if we wanted to, we weren't going to be able to shape what kind of man he would grow into.

And now that transition to manhood is getting very close. One of my favorite things (OK, not always top of *his* list) is to talk with Griffin about the powerful and fascinating changes that are just starting to take place in his body and mind—the hormone-driven changes preprogrammed by eons of evolution. One thing I can do with my knowledge of testosterone is help him to understand that his feelings throughout this transition will probably be of a different character from those of his female counterparts, and it's OK. Characteristically masculine feelings are not toxic; *he* is not toxic for having them. What matters are actions, and he has control over those. We'll try to guide him in learning to make the best, most respectful and compassionate choices. He has taught me about ways to be a boy, and he will teach me about ways to be a man. My hope is that Griffin will live in a world where men and women have complete freedom to choose careers or lifestyles, no matter how stereotypically male or female. He could be a dancer, an engineer, a nurse, a primary school teacher, or a stay-at-home dad; he could paint his toenails or take up mixed martial arts (these are not mutually exclusive).

Because of the testosterone that he is on the way to producing, Griffin will likely differ from most women in many of the ways I've described in this book. Becoming a man is a beautiful thing. But— like every man—my son should enjoy his T responsibly.

NOTES

CHAPTER 1

1   the gland decreases its production of the "sleep hormone" melatonin: R. J. Nelson and L. J. Kriegsfeld, *An Introduction to Behavioral Endocrinology*, 5th ed. (Sunderland, MA: Sinauer Associates, 2017), 73–74, 554, 703.

2   I did that with the help of a long stick: Martin N. Muller and Richard W. Wrangham, "Dominance, Aggression and Testosterone in Wild Chimpanzees: A Test of the 'Challenge Hypothesis,'" *Animal Behaviour* 67, no. 1 (2004): 113–23; 116 for methods of urine collection, preservation, and hormone analysis.

5   "Wife Beaters of Kibale": Eugene Linden, "The Wife Beaters of Kibale," *Time*, August 19, 2002, 56; see also Eugene Linden, *The Octopus and the Orangutan: More True Tales of Animal Intrigue, Intelligence, and Ingenuity* (New York: E. P. Dutton, 2002), 112.

6   *Demonic Males: Apes and the Origins of Human Violence*: Richard W. Wrangham and Dale Peterson, *Demonic Males: Apes and the Origins of Human Violence* (Boston: Houghton Mifflin Harcourt, 1996).

7   Males tend to target females: Martin N. Muller, Sonya M. Kahlenberg, Melissa Emery Thompson, and Richard W. Wrangham, "Male Coercion and the Costs of Promiscuous Mating for Female Chimpanzees," *Proceedings of the Royal Society B: Biological Sciences* 274, no. 1612 (2007): 1009–14, and Joseph T. Feldblum, Emily E. Wroblewski, Rebecca S. Rudicell, Beatrice H. Hahn, Thais Paiva, Mine Cetinkaya-Rundel, Anne E. Pusey, and Ian C. Gilby, "Sexually Coercive Male Chimpanzees Sire More Offspring," *Current Biology* 24, no. 23 (2014): 2855–60.

8   I listened carefully for mention of the Congolese rebels: Human Rights Watch, "Human Rights Watch World Report 2000—Uganda," December 1, 1999, https://www.refworld.org/docid/3ae6a8c924.html.

9   At least one woman showed evidence of a severe sexual assault: Neil Mac-
       Farquhar, "8 Tourists Slain in Uganda, Including U.S. Couple," *New York Times*,
       March 3, 1999.

9   "good numbers": Danielle Kurtzleben, "Trump and the Testosterone Take-
       over of 2016," National Public Radio, October 1, 2016, https://www.npr.org
       /2016/10/01/494249104/trump-and-the-testosterone-takeover-of-2016.

10   "rush of energy, strength, clarity, ambition, drive, impatience and, above all,
       horniness": Andrew Sullivan, "#MeToo and the Taboo Topic of Nature," *New
       York Magazine*, January 19, 2018.

10   "women are attracted to toxic masculine male phenotypes": Gad Saad, "Is
       Toxic Masculinity a Valid Concept?," *Psychology Today* blog, March 8, 2018,
       https://www.psychologytoday.com/us/blog/homo-consumericus/201803/is
       -toxic-masculinity-valid-concept.

10   Trump's presidency is "testosterone-fueled": Neal Gabler, "The Testosterone
       Fueled Presidency," *Huffington Post*, August 16, 2017, https://www.huffpost.com
       /entry/the-testosterone-fueled-presidency_b_59949cd3e4b056a2b0ef029c.

10   Trump's base was "fighting a war": Emerald Robinson, "The Collapse of the
       Never-Trump Conservatives," *American Spectator*, June 29, 2018.

10   "testosterone curse": Leon Seltzer, "Male Sexual Misconduct and the Tes-
       tosterone Curse," *Psychology Today* blog, November 29, 2017, https://www
       .psychologytoday.com/us/blog/evolution-the-self/201711/male-sexual
       -misconduct-and-the-testosterone-curse.

11   they are far more physically aggressive than women in every pocket of the
       earth, at every age: Rachel E. Morgan and Barbara A. Oudekerk, "Criminal
       Victimization, 2018," BCJ 253043, Bureau of Justice Statistics, U.S. Depart-
       ment of Justice, September 2019, https://www.bjs.gov/content/pub/pdf/cv18
       .pdf; David C. Geary, *Male, Female: The Evolution of Human Sex Differences*,
       3rd ed. (Washington, DC: American Psychological Association, 2021), 433–37;
       National Highway Traffic Safety Administration, "Comparison of Crash Fatal-
       ities by Sex and Age Group" (Washington, DC: National Center for Statis-
       tics and Analysis, 2008); Monica Hesse, "We Need to Talk About Why Mass
       Shooters Are Almost Always Men," *Washington Post*, August 5, 2019.

13   A sex difference can be a difference in the average only: there are almost
       no traits in which males do not show higher variability than females, but
       reading skill in primary school is one area in which the difference in variabil-
       ity is sometimes low and statistically non-significant. See Ariane Baye and
       Christian Monseur, "Gender Differences in Variability and Extreme Scores
       in an International Context," *Large-Scale Assessments in Education* 4, no. 1
       (2016): 1–16. For a review of sex differences in the variability of IQ and other
       cognitive tests, see Alan Feingold, "Sex Differences in Variability in Intellec-
       tual Abilities: A New Look at an Old Controversy," *Review of Educational
       Research* 62, no. 1 (1992): 61–84.

14   "denaturalizing violence": Rebecca M. Jordan-Young and Katrina Karkazis,
       Testosterone: An Unauthorized Biography (Cambridge, MA: Harvard Uni-
       versity Press, 2019), 54; and Rebecca M. Jordan-Young, "How to Kill the

'Zombie Fact' That Testosterone Drives Human Aggression," paper presented at the Women in the World: Time for a New Paradigm for Peace conference, University of Maryland, September 2019, 22.

14   As the American Psychological Association puts it: American Psychological Association, "Harmful Masculinity and Violence," *In the Public Interest* Newsletter, September 2018, https://www.apa.org/pi/about/newsletter/2018/09/harmful-masculinity.

15   "Hey Skinny": "Hey Skinny! . . . Yer Ribs Are Showing!" Digital image, The Official Website of: Charles Atlas. Charles Atlas LTD. Accessed January 15, 2020. The advertisements of Charles Atlas have been the subject of much scholarly analysis, which examines the ways in which they have reflected and promoted masculine ideals. For example, see Jacqueline Reich, "'The World's Most Perfectly Developed Man': Charles Atlas, Physical Culture, and the Inscription of American Masculinity," *Men and Masculinities* 12, no. 4 (2010): 444–61.

15   Thornhill speculated about the origins of rape in humans: Randy Thornhill, "Rape in *Panorpa* Scorpionflies and a General Rape Hypothesis," *Animal Behaviour* 28, no. 1 (1980): 52–59. Men are bigger than women, in Thornhill's words, because "larger males were favoured because of the increased likelihood of successful rape if they failed to compete successfully for parental resources" (57). For a full treatment of Thornhill and Palmer's rape hypothesis, see: Randy Thornhill and Craig T. Palmer, *A Natural History of Rape: Biological Bases of Sexual Coercion* (Cambridge, MA: MIT Press, 2001). For a critical review, see Jerry A. Coyne and Andrew Berry, "Rape as an Adaptation," *Nature* 404, no. 6774 (2000): 121–22.

16   emotional reactions, whether positive or negative: Justin Storbeck and Gerald L. Clore, "Affective Arousal as Information: How Affective Arousal Influences Judgments, Learning, and Memory," *Social and Personality Psychology Compass* 2, no. 5 (2008): 1824–43.

18   "So my best guess, to provoke you": Lawrence Summers, "Full Transcript: President Summers' Remarks at the National Bureau of Economic Research, Jan. 14 2005," *Harvard Crimson*, February 18, 2005, https://www.thecrimson.com/article/2005/2/18/full-transcript-president-summers-remarks-at/.

18   Summers resigned: Alan Finder, "President of Harvard Resigns, Ending Stormy 5-Year Tenure," *New York Times*, February 22, 2006.

19   "We've trained young men to be adventurous": Sara Rimer and Patrick D. Healy, "Furor Lingers as Harvard Chief Gives Details of Talk on Women," *New York Times*, February 18, 2005.

20   "the distribution of preferences and abilities of men and women differ": James Damore, "Google's Ideological Echo Chamber," July 2017, https://assets.documentcloud.org/documents/3914586/Googles-Ideological-Echo-Chamber.pdf.

20   Some cognitive scientists who reviewed his claims found them to be supported by the evidence: See the contributions by Lee Jussim, Geoffrey Miller, and Debra W. Soh in "The Google Memo: Four Scientists Respond,"

Quillette, August 17, 2017, https://quillette.com/2017/08/07/google-memo-four-scientists-respond/; Debra Soh, "No, the Google Manifesto Isn't Sexist or Anti-Diversity. It's Science," *Globe and Mail* (Toronto), August 8, 2017; Glenn Stanton, "The Science Says the Google Guy Was Right About Sex Differences," *Federalist*, August 11, 2017, https://thefederalist.com/2017/08/11/science-says-google-guy-right-sex-differences/. For more critical views, see Megan Molteni and Adam Rogers, "The Actual Science of James Damore's Google Memo," *Wired*, August 15, 2017, https://www.wired.com/story/the-pernicious-science-of-james-damores-google-memo/; and Brian Feldman, "Here Are Some Scientific Arguments James Damore Has Yet to Respond To," *New York Magazine*, August 11, 2017.

20 others were more critical: Gina Rippon, "What Neuroscience Can Tell Us About the Google Diversity Memo," *The Conversation*, August 14, 2017, https://theconversation.com/what-neuroscience-can-tell-us-about-the-google-diversity-memo-82455. Rippon characterizes Damore's memo as claiming that "women are underrepresented in the tech industry because of biological differences," but Damore never said that biology was the entire explanation. Rippon also notes that sex differences in interests and aptitudes do not "fall into two neat categories" and "exist on a spectrum," as if this is an objection. Again, Damore never claimed otherwise.

20 nor did it prevent Google from firing Damore: Daisuke Wakabayashi, "Contentious Memo Strikes Nerve Inside Google and Out," *New York Times*, August 8, 2017.

20 This was no doubt unfortunate for Damore, who subsequently sued Google: Daisuke Wakabayashi and Nellie Bowles, "Google Memo Author Sues, Claiming Bias Against White Conservative Men," *New York Times*, January 8, 2018.

21 alleged biological basis of women's inferiority: Angela Saini, *Inferior: How Science Got Women Wrong and the New Research That's Rewriting the Story* (Boston: Beacon Press, 2017).

21 "the average of mental power in man must be above that of woman": Charles Darwin, *The Descent of Man, and Selection in Relation to Sex*, 2 vols. (London: John Murray, 1871), vol. 1, 564.

22 women have overtaken men: David F. Feldon, James Peugh, Michelle A. Maher, Josipa Roksa, and Colby Tofel-Grehl, "Time-to-Credit Gender Inequities of First-Year PhD Students in the Biological Sciences," *CBE—Life Sciences Education* 16, no. 1 (2017), article 4. As of 2017, women earned 52.5 percent of PhD degrees in the biological sciences but held only 30 to 35 percent of tenure-track positions in related fields.

22 "bringing forth of strong children": Gertrud Pfister, "The Medical Discourse on Female Physical Culture in Germany in the 19th and Early 20th Centuries," *Journal of Sport History* 17, no. 2 (1990): 191.

22 Before the eugenics laws were abolished: Adam S. Cohen, "Harvard's Eugenics Era," *Harvard Magazine*, March–April 2016, https://harvardmagazine.com/2016/03/harvards-eugenics-era.

23   "squashes the hopes for sex equality": Publisher's description, back cover of
      Cordelia Fine, *Testosterone Rex: Myths of Sex, Science, and Society* (London:
      Icon Books, 2017).
23   you're falling for an "overly familiar story": Publisher's description, W. W.
      Norton, Fine, *Testosterone Rex*, https://www.wwnorton.co.uk/books
      /9780393082081-testosterone-rex.
24   "This book brilliantly explains": "Cordelia Fine's Explosive Study of Gender
      Politics Wins 30th Anniversary Royal Society Insight Investment Science
      Book Prize," Royal Society, news, September 19, 2017, https://royalsociety
      .org/news/2017/09/cordelia-fine-wins-30th-anniversary-royal-society
      -insight-investment-science-book-prize/.
24   "ingrained myths": Fine, *Testosterone Rex*, publisher's description.
24   "A belief in biology brings with it a particular mindset": Gina Rippon, *The
      Gendered Brain: The New Neuroscience That Shatters the Myth of the Female
      Brain* (New York: Random House, 2019), 353.
24   "Does the balance of sex hormones": Saini, *Inferior*, 28.
24   "There are few psychological differences between the sexes": Saini, *Inferior*, 10.
25   social progress depends on scientific progress: For a book-length treatment of
      how science and reason promote social progress, see Steven Pinker, *Enlight-
      enment Now: The Case for Reason, Science, Humanism, and Progress* (New
      York: Viking, 2018).

CHAPTER 2

27   genes for internal testes: Virag Sharma, Nikolai Hecker, Juliana G. Roscito,
      Leo Fourster, Bjoern I. Langer, and Michael Hiller, "A Genomics Approach
      Reveals Insights into the Importance of Gene Losses for Mammalian Adap-
      tations," *Nature Communications* 9, no. 1 (2018): 1215–19.
27   stay away from tighty whities: Damayanthi Durairajanayagam, Rakesh K.
      Sharma, Stefan S. du Plessis, and Ashok Agarwal, "Testicular Heat Stress and
      Sperm Quality," in *Male Infertility*, ed. Stefan S. du Plessis, Ashok Agarwal,
      and Edmund S. Sabanegh Jr. (New York: Springer, 2014), 105–25.
28   mammals with internal testes: Sharma et al., "A Genomics Approach Reveals
      Insights."
28   "a eunuch never goes bald": Aristotle, "History of Animals," in *Complete
      Works of Aristotle*, vol. 1, rev. Oxford translation, ed. Jonathan Barnes (Prince-
      ton, NJ: Princeton University Press, 1984), 981.
29   The article quoted the famed Italian opera singer Cecilia Bartoli: Angela
      Giuffrida, "Sistine Chapel Breaks 500-Year Gender Taboo to Welcome
      Soprano into the Choir," *Guardian*, November 18, 2017.
30   In spite of the risks: Meyer M. Melicow, "Castrati Singers and the Lost
      'Cords,'" *Bulletin of the New York Academy of Medicine* 59, no. 8 (1983):
      744.
31   wielded significant political power: Robert B. Crawford, "Eunuch Power in
      the Ming Dynasty," *T'oung Pao* 49, no. 1 (1962): 115–48; Eberhard Nieschlag

and Susan Nieschlag, "The History of Testosterone and the Testes: From Antiquity to Modern Times," in Lee B. Smith, *Testosterone: From Basic Research to Clinical Applications* (New York: Springer, 2017), 1–19.

31   "There can be no other motive in using them": G. Carter Stent, "Chinese Eunuchs," *Journal of North-China Branch of the Royal Asiatic Society* 10 (1877): 143.

32   "Being a cripple is better than starving": Yinghua Jia, *The Last Eunuch of China: The Life of Sun Yaoting* (Beijing: China Intercontinental Press, 2008), 8.

33   preserved organs: Jean D. Wilson and Claus Roehrborn, "Long-Term Consequences of Castration in Men: Lessons from the Skoptzy and the Eunuchs of the Chinese and Ottoman Courts," *Journal of Clinical Endocrinology and Metabolism* 84, no. 12 (1999): 4324–31. Chinese eunuchs traditionally had both their testicles and penis removed, but whether the penis was removed along with the testicles varied across time and culture; see Kathryn M. Ringrose, "Eunuchs in Historical Perspective," *History Compass* 5, no. 2 (2007): 495–506.

33   "nothing can save him": Stent, "Chinese Eunuchs," 177.

33   "I have tasted both joys and sorrows": Jia, *Last Eunuch of China*, 8.

36   would they somehow transform the blood: Lynn Loriaux, *A Biographical History of Endocrinology* (Ames, IA: Wiley-Blackwell, 2016).

36   "So far as voice, sexual urge, belligerence, and growth of comb and wattles are concerned": Arnold Adolph Berthold, "The Transplantation of Testes," trans. D. P. Quiring, *Bulletin of the History of Medicine* 16, no. 4 (1944): 399–401, 401.

37   *"by their action on the blood stream"*: Berthold, "Transplantation of Testes," 401 (my emphasis).

37   Berthold was not the first to postulate: John Hunter, a Scottish surgeon, reported results similar to those of Berthold in some of his lectures, but he never published any studies on these observations: Alvaro Morales, "The Long and Tortuous History of the Discovery of Testosterone and Its Clinical Application," *Journal of Sexual Medicine* 10, no. 4 (2013): 1178–93; Garabed Eknoyan, "Emergence of the Concept of Endocrine Function and Endocrinology," *Advances in Chronic Kidney Disease* 11, no. 4 (2004): 371–76.

38   The esteemed and influential German anatomist and zoologist Charles-Édouard Brown-Séquard: Setti S. Rengachary, Chaim Colen, and Murali Guthikonda, "Charles-Edouard Brown-Séquard: An Eccentric Genius," *Neurosurgery* 62, no. 4 (2008): 954–64.

38   The loss of semen through sex or masturbation: Merriley Borell, "Organotherapy, British Physiology, and Discovery of the Internal Secretions," *Journal of the History of Biology* 9, no. 2 (1976): 235–68.

38   "power is possessed by some substance": C. E. Brown-Séquard, "Note on the Effects Produced on Man by Subcutaneous Injections of a Liquid Obtained from the Testicles of Animals," *Lancet* 134, no. 3438 (1889): 105–107.

39   Testes from accident victims: Erica R. Freeman, David A. Bloom, and Edward J. McGuire, "A Brief History of Testosterone," *Journal of Urology* 165, no. 2 (2001): 371–73.

39 almost certainly due to the placebo effect: Andrea J. Cussons, John P. Walsh, Chotoo I. Bhagat, and Stephen J. Fletcher, "Brown-Séquard Revisited: A Lesson from History on the Placebo Effect of Androgen Treatment," *Medical Journal of Australia* 177, no. 11 (2002): 678–79.

39 The restorative powers of organ extracts: Eberhard Nieschlag and Susan Nieschlag, "The History of Discovery, Synthesis and Development of Testosterone for Clinical Use," *European Journal of Endocrinology* 180, no. 6 (2019): R201–R212.

40 sodium bicarbonate, which counteracts: J. D. Kaunitz and Y. Akiba, "Duodenal Bicarbonate: Mucosal Protection, Luminal Chemosensing and Acid-Base Balance," *Alimentary Pharmacology and Therapeutics* 24, no. s4 (2006): 169–76.

40 "we were dealing with an entirely different order of phenomena": William Maddock Bayliss and Ernest Henry Starling, "The Mechanism of Pancreatic Secretion," *Journal of Physiology* 28, no. 5 (1902): 322, 325–53.

41 Pavlov won the Nobel Prize: J. H. Henriksen and O. B. Schaffalitzky de Muckadell, "Secretin, Its Discovery, and the Introduction of the Hormone Concept," *Scandinavian Journal of Clinical and Laboratory Investigation* 60, no. 6 (2000): 463–72; "The Nobel Prize in Physiology or Medicine 1904," The Nobel Prize, https://www.nobelprize.org/prizes/medicine/1904/summary/.

41 "These chemical messengers, however, or hormones": John Henderson, "Ernest Starling and 'Hormones': An Historical Commentary," *Journal of Endocrinology* 184, no. 1 (2005): 5–10.

41 The discovery of secretin: For the history of Starling and Bayliss's discovery of the first hormone and its impact on our understanding of the regulation of internal states and processes, see Henriksen and Schaffalitzky de Muckadell, "Secretin," 463–72.

42 Butenandt was awarded the Nobel Prize: Morales, "The Long and Tortuous History of the Discovery of Testosterone and Its Clinical Application"; and Eberhard Nieschlag and Susan Nieschlag, "Testosterone Deficiency: A Historical Perspective," *Asian Journal of Andrology* 16, no. 2 (2014): 161–68.

49 being synthesized and sold for use in pharmacology: Nieschlag and Nieschlag, "Testosterone Deficiency."

49 normal development of most male animals requires testosterone exposure: Lucia Lanciotti, Marta Cofini, Alberto Leonardi, Laura Penta, and Susanna Esposito, "Up-to-Date Review About Minipuberty and Overview on Hypothalamic-Pituitary-Gonadal Axis Activation in Fetal and Neonatal Life," *Frontiers in Endocrinology* 9 (2018), article 410. Testosterone also rises in male infants shortly after birth and stays high for about three months. A similar phenomenon happens in girls, but it is estrogen that rises. This period, called "minipuberty," is increasingly appreciated as a critical period for the maturation of the reproductive organs and possibly for further sexual differentiation of the nervous system. Much remains to be learned about this period of development.

CHAPTER 3

52   Jenny had a rare DSD: Ieuan A. Hughes, John D. Davies, Trevor I. Bunch et al., "Androgen Insensitivity Syndrome," *Lancet* 380, no. 9851 (2012): 1419–28. Estimates for the prevalence of CAIS vary, from about 1 to 5 per 100,000 XY individuals.

54   You may have thought that possessing XY sex chromosomes in (nearly) every cell is incompatible: In mammals, all cells except for the gametes (eggs and sperm) and red blood cells normally contain the entire set of twenty-two pairs of autosomal chromosomes (the numbered chromosomes) and one pair of sex chromosomes, either XX or XY, with one sex chromosome donated from each parent. Meiosis is the process through which gametes are created from parent cells. In this process, maternal and paternal chromosomes in each sex cell "cross over," trading pieces of DNA, so that each chromosome in the pair contains a mix of maternal and paternal DNA. One chromosome from each pair, including the sex chromosomes, is then passed into each new egg or sperm cell. (Cells with single chromosomes instead of pairs are called "haploid cells.") When the gametes combine through fertilization, they form a new cell containing 23 pairs of chromosomes, containing one from each parent. This "diploid cell" starts rapidly dividing into a mass of diploid cells, each cell containing the DNA of both parents, forming the "blastocyst" and then the embryo.

54   But the number of genes on the Y is minute: Steven L. Salzberg, "Open Questions: How Many Genes Do We Have?," *BMC Biology* 16, no. 1 (2018), article 94. We still do not know exactly how many genes are contained in the human genome. Estimates have declined over time and most recently are in the range of twenty thousand to twenty-five thousand. The term "gene" can also refer to different things but usually means stretches of DNA that are translated into RNA and then into protein, or that are translated into RNA but not into protein. These "non-coding RNAs," or ncRNAs, have been found to regulate gene expression, and other functions are being discovered.

55   One of its genes makes all the difference: For a fascinating description of the work of David Page of MIT, arguably the top expert in the world on the Y chromosome, who has "brought dignity and respect to this chromosomal runt," see Bijal Trivedi, "Profile of David C. Page," *Proceedings of the National Academy of Sciences* 103, no. 8 (2006): 2471–73.

55   End to end, there's about six feet of it in each cell: Bruce Alberts, Alexander Johnson, Julian Lewis, Martin Raff, Keith Roberts, and Peter Walter, "Chromosomal DNA and Its Packaging in the Chromatin Fiber," in *Molecular Biology of the Cell*, 4th ed. (New York: Garland Science, 2002).

55   DNA from all of your cells could stretch to the sun and back: Melvin L. DePamphilis, Christelle M. de Renty, Zakir Ullah, and Chrissie Y. Lee, "'The Octet': Eight Protein Kinases That Control Mammalian DNA Replication," *Frontiers in Physiology* 3 (2012), article 368.

59   Of particular importance is the gene SOX9 on chromosome 17: Helena Sim, Anthony Argentaro, and Vincent R. Harley, "Boys, Girls and Shuttling of SRY and SOX9," *Trends in Endocrinology & Metabolism* 19, no. 6 (2008): 213–22.

61   But that's not how it works: For an accessible and comprehensive overview of the variety and complexity of sex determination across many different kinds of organisms, from flies to fish to ferns (and humans), see Doris Bachtrog, Judith E. Mank, Catherine L. Peichel et al., "Sex Determination: Why So Many Ways of Doing It?," *PLOS Biology* 12, no. 7 (2014): e1001899.

61   Males produce small: Joan Roughgarden, *Evolution's Rainbow: Diversity, Gender, and Sexuality in Nature and People* (Berkeley: University of California Press, 2013), 23.

61   it's the design plan for the gametes that counts: Sometimes human males don't have XY sex chromosomes, and females don't have XX. In rare cases during meiosis (the creation of sex cells, or sperm and eggs), the SRY gene may become "translocated" from the Y chromosome to another chromosome, most often the X (with which it recombines and separates during meiosis). If an X chromosome containing SRY is donated by the father, (and an X is donated by the mother, as usual), then the resulting individual is a male with XX sex chromosomes, because the SRY gene leads to the development of testes rather than ovaries. These XX males are usually infertile because they lack genes on the Y-chromosome that are necessary for full reproductive function. See Ahmad Majzoub, Mohamed Arafa, Christopher Starks, Haitham Elbardisi, Sami Al Said, and Edmund Sabanegh, "46 XX Karyotype During Male Fertility Evaluation: Case Series and Literature Review," *Asian Journal of Andrology* 19, no. 2 (March–April 2017): 168–72.

65   These sorts of mutations produce partial androgen insensitivity syndrome: For more on CAIS and PAIS, see Hughes et al., "Androgen Insensitivity Syndrome."

65   Only some genes are responsive to sex steroids: Thomas M. Williams and Sean B. Carroll, "Genetic and Molecular Insights into the Development and Evolution of Sexual Dimorphism," *Nature Reviews Genetics* 10, no. 11 (2009): 797–804; Cho-Yi Chen, Camila Lopes-Ramos, Marieke L. Kuijjer, Joseph N. Paulson, Abhijeet R. Sonawane, Maud Fagny, John Platig et al., "Sexual Dimorphism in Gene Expression and Regulatory Networks Across Human Tissues," *BioRxiv* (2016): 082289.

68   Bodybuilders are very aware of this: Shehzad Basaria, "Androgen Abuse in Athletes: Detection and Consequences," *Journal of Clinical Endocrinology and Metabolism* 95, no. 4 (2010): 1533–43.

69   she chose to keep her internal testicles: The decision about whether to retain one's testes in CAIS is not easy. Historically, many doctors recommended the removal of the internal testes in CAIS women, due to an increased risk of the development of cancerous tumors in the testes. But this practice is now questioned, since many women with CAIS would rather avoid surgery and get the hormones they need (mainly estrogen) naturally, from their

own bodies, rather than being dependent on exogenous hormones. CAIS women who choose this path are advised to have regular monitoring of their internal testes to ensure they stay healthy. See M. Cools and L. Looijenga, "Update on the Pathophysiology and Risk Factors for the Development of Malignant Testicular Germ Cell Tumors in Complete Androgen Insensitivity Syndrome," *Sexual Development* 11, no. 4 (2017): 175–81; and U. Döhnert, L. Wünsch, and O. Hiort, "Gonadectomy in Complete Androgen Insensitivity Syndrome: Why and When?," *Sexual Development* 11, no. 4 (2017): 171–74.

69   Jenny's T was impotent as an agent of masculinization: You may be wondering why men's T doesn't get converted into estrogen and feminize them. CAIS women have estrogen levels that are in the normal male range, significantly lower than normal female levels. Yet the reason that these low levels of estrogen can feminize CAIS women, and not typical males, is because in CAIS, estrogen can act completely unopposed by androgens. High levels of androgens normally swamp the effects of low levels of estrogen, preventing the development of feminine physical features. See U. Doehnert, S. Bertelloni, R. Werner, E. Dati, and O. Hiort, "Characteristic Features of Reproductive Hormone Profiles in Late Adolescent and Adult Females with Complete Androgen Insensitivity Syndrome," *Sexual Development* 9, no. 2 (2015): 69–74; and Dimitrios T. Papadimitriou, Agnès Linglart, Yves Morel, and Jean-Louis Chaussain, "Puberty in Subjects with Complete Androgen Insensitivity Syndrome," *Hormone Research in Paediatrics* 65, no. 3 (2006): 126–31.

71   Anne Fausto-Sterling describes a study called "Baby X Revisited": Anne Fausto-Sterling, *Myths of Gender: Biological Theories About Women and Men*, rev. ed. (New York: Basic Books, 2008), 137.

71   girls' play . . . focuses on coming together: For sex differences in play, particularly boys' obsession with enemies, see Joyce F. Benenson, *Warriors and Worriers: The Survival of the Sexes* (New York: Oxford University Press, 2014), 27–40; and for a comprehensive review of sex differences in play, see David C. Geary, *Male, Female: The Evolution of Human Sex Differences*, 3rd ed. (Washington, DC: American Psychological Association, 2021), 309–23.

72   Boys, it seems, are resistant: Benenson, *Warriors and Worriers*, 27–41.

72   Boys tend to prefer this "rough-and-tumble" play: Janet A. DiPietro, "Rough and Tumble Play: A Function of Gender," *Developmental Psychology* 17, no. 1 (1981): 50–58; Anthony D. Pellegrini, "The Development and Function of Rough-and-Tumble Play in Childhood and Adolescence: A Sexual Selection Theory Perspective," in *Play and Development: Evolutionary, Sociocultural and Functional Perspectives*, ed. Artin Göncü and Suzanne Gaskins (Mahwah, NJ: Lawrence Erlbaum, 2007); Yumi Gosso, Emma Otta, and Maria de Lima Salum e Morais, "Play in Hunter-Gatherer Society," in *The Nature of Play: Great Apes and Humans*, ed. Anthony D. Pellegrini and Peter K. Smith (New York: Guilford Press, 2004), 231; David C. Geary, "Evolution and Developmental Sex Differences," *Current Directions in Psychological Science* 8, no. 4 (1999): 115–20; and Sheina Lew-Levy, Adam H. Boyette, Alyssa N. Crittenden, Barry S. Hewlett, and Michael E. Lamb, "Gender-Typed and

Gender-Segregated Play Among Tanzanian Hadza and Congolese BaYaka Hunter-Gatherer Children and Adolescents," *Child Development* 91, no. 4 (2020): 1284–301.

73   Fausto-Sterling uses the Baby X experiment: Fausto-Sterling, *Myths of Gender*, 137.

74   In her 2010 award-winning book *Brain Storm*: R. M. Jordan-Young, *Brain Storm: The Flaws in the Science of Sex Differences* (Cambridge, MA: Harvard University Press, 2011), 291.

74   "A gendered world will produce a gendered brain": Gina Rippon, *The Gendered Brain: The New Neuroscience That Shatters the Myth of the Female Brain* (New York: Random House, 2019), xix.

74   "neurosexism": Lise Eliot, "Neurosexism: The Myth That Men and Women Have Different Brains," *Nature* 566, no. 7745 (2019): 453–54.

CHAPTER 4

75   Clad in the traditional hijab and long dress: Diana Mettadewi Jong, Aman B. Pulungan, Bambang Tridjaja Aap, and Jose R. L. Batubara, "5-alpha-reductase Deficiency: A Case Report," *Paediatrica Indonesiana* 43, no. 6 (2003): 234–40.

76   he developed external genitalia that appeared female: Berenice B. Mendonca, Rafael Loch Batista, Sorahia Domenice, Elaine M. F. Costa, Ivo J. P. Arnhold, David W. Russell, and Jean D. Wilson, "Steroid 5α-Reductase 2 Deficiency," *Journal of Steroid Biochemistry and Molecular Biology* 163 (2016): 206–11.

76   This extra stimulation comes from the more potent androgen dihydrotestosterone (DHT): For an overview of the roles of T and DHT in 5-ARD, see Julianne Imperato-McGinley and Y.-S. Zhu, "Androgens and Male Physiology the Syndrome of 5α-Reductase-2 Deficiency," *Molecular and Cellular Endocrinology* 198, no. 1–2 (2002): 51–59.

77   High levels of T can do a decent job on their own: John C. Achermann and Ieuan A. Hughes, "Pediatric Disorders of Sex Development," in *Williams Textbook of Endocrinology*, 13th ed., ed. Shlomo Melmed, Kenneth Polonsky, P. Larsen, and Henry Kronenberg (Philadelphia: Elsevier Health Sciences, 2016), ch. 23.

78   boys were given wider latitude to "romp and play": Julianne Imperato-McGinley, Ralph E. Peterson, Teofilo Gautier, and Erasmo Sturla, "Androgens and the Evolution of Male-Gender Identity Among Male Pseudohermaphrodites with 5α-Reductase Deficiency," *New England Journal of Medicine* 300, no. 22 (1979): 1236.

79   While the older boys and young men attended cockfights: Imperato-McGinley et al., "Androgens and the Evolution of Male-Gender Identity," 1237.

79   "Their dawning discovery of maleness": Vivian Sobel and Julianne Imperato-McGinley, "Gender Identity in XY Intersexuality," *Child and Adolescent Psychiatric Clinics of North America* 13, no. 3 (2004): 611.

80   But these proportions vary: Peggy T. Cohen-Kettenis, "Gender Change in 46 XY Persons with 5α-Reductase-2 Deficiency and 17β-Hydroxysteroid

Dehydrogenase-3 Deficiency," *Archives of Sexual Behavior* 34, no. 4 (2005): 399–410; and Rafael Loch Batista and Berenice Bilharinho Mendonca, "Integrative and Analytical Review of the 5-Alpha-Reductase Type 2 Deficiency Worldwide," *Application of Clinical Genetics* 13 (2020): 83–96.

80   "a masculine identity was evolving": Ruth Bleier, J. A. Keelan, Julianne Imperato-McGinley, and Ralph E. Peterson, "Why Does a Pseudohermaphrodite Want to Be a Man?," correspondence, *New England Journal of Medicine* 301, no. 15 (1979): 839–40.

80   "When she turned five I noticed": BBC, "The Extraordinary Case of the Guevedoces," *BBC News Magazine*, September 15, 2015, https://www.bbc.com /news/magazine-34290981.

80   her article has been cited nearly fifteen hundred times: As of December 2020, Julianne Imperato-McGinley, Luis Guerrero, Teofilo Gautier, and Ralph E. Peterson, "Steroid 5α-Reductase Deficiency in Man: An Inherited Form of Male Pseudohermaphroditism," *Science* 186, no. 4170 (1974): 1212–15, has been cited 1,488 times, according to Google Scholar.

81   "in addition to behavioral differences": Imperato-McGinley et al., "Androgens and the Evolution of Male-Gender Identity," 1235.

81   "faulty logic and restrictive interpretations": Bleier et al., "Why Does a Pseudohermaphrodite Want to Be a Man?," 840.

81   "women's subordinate position inevitable": Ruth Bleier, *Science and Gender: A Critique of Biology and Its Theories on Women* (New York: Pergamon Press, 1984), 109.

82   She's motivated, but he still needs to put in some effort: Frank A. Beach, "Sexual Attractivity, Proceptivity, and Receptivity in Female Mammals," *Hormones and Behavior* 7, no. 1 (1976): 105–38.

82   Lordosis is a common mating pose: R. J. Nelson and L. J. Kriegsfeld, *An Introduction to Behavioral Endocrinology*, 5th ed. (Sunderland, MA: Sinauer Associates, 2017), 283–84.

83   once-sexy males lose their luster, and she also loses her reflexive lordosis pose: For a review of the role of hormones in "programming" lordosis and mounting, see Arthur P. Arnold, "The Organizational-Activational Hypothesis as the Foundation for a Unified Theory of Sexual Differentiation of All Mammalian Tissues," *Hormones and Behavior* 55, no. 5 (2009): 570–78.

83   gonadal hormones clearly drive the behaviors necessary to reproduce: Nelson and Kriegsfeld, *Introduction to Behavioral Endocrinology*, 216–22.

83   One couldn't simply get a female to mount: Nelson and Kriegsfeld, *Introduction to Behavioral Endocrinology*, 120–21.

84   "Counteracting this deterrent": William C. Young, Robert W. Goy, and Charles H. Phoenix, "Hormones and Sexual Behavior," *Science* 143, no. 3603 (1964): 212–18 (my emphasis).

85   the second, adult dose would activate: Charles H. Phoenix, Robert W. Goy, Arnold A. Gerall, and William C. Young, "Organizing Action of Prenatally Administered Testosterone Propionate on the Tissues Mediating Mating Behavior in the Female Guinea Pig," *Endocrinology* 65, no. 3 (1959): 369–82.

86   Fetal exposure to high T had squashed the capacity: Arnold, "The Organizational-Activational Hypothesis as the Foundation for a Unified Theory of Sexual Differentiation of All Mammalian Tissues."

86   If the brain is not masculinized prenatally: Phoenix et al., "Organizing Action of Prenatally Administered Testosterone Propionate."

86   "evidence that the rodent model does not apply to primates or humans": Bleier et al., "Why Does a Pseudohermaphrodite Want to Be a Man?," 840.

86   the basic results of Young's paper were replicated in a study of rhesus macaque monkeys: R. W. Goy and J. A. Resko, "Gonadal Hormones and Behavior of Normal and Pseudohermaphroditic Nonhuman Female Primates," *Recent Progress in Hormone Research* 28 (1972): 707–33.

87   play behavior increases reproductive success: That's one version of the "proximate/ultimate" distinction; in another version, proximate explanations cite causes that occur immediately before the phenomenon to be explained, while ultimate explanations cite causes that occur in the distant past. These two versions are easy to confuse, but they are different. See David Haig, "Proximate and Ultimate Causes: How Come? and What For?," *Biology and Philosophy* 28, no. 5 (2013): 781–86.

87   Dominant males who can win fights mate more: D. H. Thor and W. J. Carr, "Sex and Aggression: Competitive Mating Strategy in the Male Rat," *Behavioral and Neural Biology* 26, no. 3 (1979): 261–65.

87   women also form dominance hierarchies: Anne Campbell, "Staying Alive: Evolution, Culture, and Women's Intrasexual Aggression," *Behavioral and Brain Sciences* 22, no. 2 (1999): 203–14.

88   "Boxing occurs when both rats stand on their hindpaws": Anthony P. Auger and Kristin M. Olesen, "Brain Sex Differences and the Organization of Juvenile Social Play Behavior," *Journal of Neuroendocrinology* 21, no. 6 (2009): 519–25.

88   They are poor fighters who readily submit to intruders: Dorothy Einon and Michael Potegal, "Enhanced Defense in Adult Rats Deprived of Playfighting Experience as Juveniles," *Aggressive Behavior* 17, no. 1 (1991): 27–40; and Aileen D. Gruendel and William J. Arnold, "Influence of Preadolescent Experiential Factors on the Development of Sexual Behavior in Albino Rats," *Journal of Comparative and Physiological Psychology* 86, no. 1 (1974): 172–78.

88   amount of T in the babies' circulation affects how much mom licks and grooms: Celia Moore, "Maternal Behavior of Rats Is Affected by Hormonal Condition of Pups," *Journal of Comparative and Physiological Psychology* 96, no. 1 (1982): 123–29.

88   they need a longer recovery period before they can ejaculate again: Celia L. Moore, "Maternal Contributions to the Development of Masculine Sexual Behavior in Laboratory Rats," *Developmental Psychobiology* 17, no. 4 (1984): 347–56; and Lynda I. A. Birke and Dawn Sadler, "Differences in Maternal Behavior of Rats and the Sociosexual Development of the Offspring," *Developmental Psychobiology* 20, no. 1 (1987): 85–99.

89   according to newer evidence, also in the first few months after birth:
      Annamarja Lamminmäki, Melissa Hines, Tanja Kuiri-Hänninen et al., "Tes-
      tosterone Measured in Infancy Predicts Subsequent Sex-Typed Behavior in
      Boys and in Girls," *Hormones and Behavior* 61, no. 4 (2012): 611–16.

90   "Identification of common neurochemical and neuroanatomical substrates
      of sexual responding": James G. Pfaus, Tod E. Kippin, and Genaro Coria-Avila,
      "What Can Animal Models Tell Us About Human Sexual Response?," *Annual
      Review of Sex Research* 14, no. 1 (2003): 1–63.

91   Congenital adrenal hyperplasia (CAH) is a rare genetic disorder: R. Schweizer,
      G. Blumenstock, K. Mangelsdorf et al., "Prevalence and Incidence of Endo-
      crine Disorders in Children: Results of a Survey in Baden-Wuerttemberg and
      Bavaria (EndoPrIn BB) 2000–2001," *Klinische Pädiatrie* 222, no. 2 (2010):
      67–72; P. W. Speiser, W. Arlt, R. J. Auchus, L. S. Baskin, G. S. Conway, D. P.
      Merke, H. F. L. Meyer-Bahlburg et al., "Congenital Adrenal Hyperplasia Due
      to Steroid 21-Hydroxylase Deficiency: An Endocrine Society Clinical Prac-
      tice Guideline," *Journal of Clinical Endocrinology and Metabolism* 103, no. 11
      (2018): 4043–88.

91   but significantly affects the behavior of girls only: CAH in males does not
      appear to affect gendered behavior, such as play style, sexual orientation, or
      later career choices, but it may affect aspects of male cognition, in partic-
      ular decreasing spatial ability (CAH does not appear to affect spatial abil-
      ity in females). See Marcia L. Collaer and Melissa Hines, "No Evidence for
      Enhancement of Spatial Ability with Elevated Prenatal Androgen Exposure
      in Congenital Adrenal Hyperplasia: A Meta-Analysis," *Archives of Sexual
      Behavior* 49, no. 2 (2020): 395–411.

91   opportunity to investigate the effects of early androgen exposure on the devel-
      oping brain in humans: Sheri A. Berenbaum and Adriene M. Beltz, "Sexual
      Differentiation of Human Behavior: Effects of Prenatal and Pubertal Organiza-
      tional Hormones," *Frontiers in Neuroendocrinology* 32, no. 2 (2011): 183–200.

92   these girls look typically female: Rafał Podgórski, David Aebisher, Mon-
      ika Stompor, Dominika Podgórska, and Artur Mazur, "Congenital Adrenal
      Hyperplasia: Clinical Symptoms and Diagnostic Methods," *Acta Biochimica
      Polonica* 65, no. 1 (2018): 25–33.

93   one classic experiment: William R. Charlesworth and Claire Dzur, "Gender
      Comparisons of Preschoolers' Behavior and Resource Utilization in Group
      Problem Solving," *Child Development* 58, no. 1 (1987): 191–200.

94   boys are far more likely to compete physically: Joyce F. Benenson, *Warriors
      and Worriers: The Survival of the Sexes* (New York: Oxford University Press,
      2014), 45–51; and Amanda J. Rose and Karen D. Rudolph, "A Review of Sex
      Differences in Peer Relationship Processes: Potential Trade-offs for the Emo-
      tional and Behavioral Development of Girls and Boys," *Psychological Bulletin*
      132, no. 1 (2006): 98–131.

94   The overwhelming majority of children's playmates are members of their own
      sex: Eleanor E. Maccoby, *The Two Sexes: Growing Up Apart, Coming Together*
      (Cambridge, MA: Harvard University Press, 1999): 27; Joyce F. Benenson,

Nicholas H. Apostoleris, and Jodi Parnass, "Age and Sex Differences in Dyadic and Group Interaction," *Developmental Psychology* 33, no. 3 (1997): 538–43.

94   boys have more to lose: Beverly I. Fagot, "Consequences of Moderate Cross-Gender Behavior in Preschool Children," *Child Development* 48, no. 3 (1977): 902–7.

94   Sex differences in children's play presage: Elizabeth V. Lonsdorf, "Sex Differences in Nonhuman Primate Behavioral Development," *Journal of Neuroscience Research* 95, no. 1–2 (2017): 213–21; Joyce F. Benenson, "Sex Differences in Human Peer Relationships: A Primate's-Eye View," *Current Directions in Psychological Science* 28, no. 2 (2019): 124–30; Janice M. Hassett, Erin R. Siebert, and Kim Wallen, "Sex Differences in Rhesus Monkey Toy Preferences Parallel Those of Children," *Hormones and Behavior* 54, no. 3 (2008): 359–64; Beatrice Whiting and Carolyn Pope Edwards, "A Cross-Cultural Analysis of Sex Differences in the Behavior of Children Aged Three Through 11," *Journal of Social Psychology* 91, no. 2 (1973): 171–88; and Jac T. M. Davis and Melissa Hines, "How Large Are Gender Differences in Toy Preferences? A Systematic Review and Meta-Analysis of Toy Preference Research," *Archives of Sexual Behavior* 49, no. 2 (2020): 373–94.

94   aggression, parenting, social hierarchies, and preferences: Rong Su, James Rounds, and Patrick Ian Armstrong, "Men and Things, Women and People: A Meta-Analysis of Sex Differences in Interests," *Psychological Bulletin* 135, no. 6 (2009): 859–84.

95   The "girls' toys" included a cosmetics kit: Vickie L. Pasterski, Mitchell E. Geffner, Caroline Brain, Peter Hindmarsh, Charles Brook, and Melissa Hines, "Prenatal Hormones and Postnatal Socialization by Parents as Determinants of Male-Typical Toy Play in Girls with Congenital Adrenal Hyperplasia," *Child Development* 76, no. 1 (2005): 264–78.

96   These studies confirm: Sheri A. Berenbaum and Adriene M. Beltz, "Evidence and Implications from a Natural Experiment of Prenatal Androgen Effects on Gendered Behavior," *Current Directions in Psychological Science* 30, no. 3 (2021): 202–10; Melissa Hines, Vickie Pasterski, Debra Spencer, Sharon Neufeld, Praveetha Patalay, Peter C. Hindmarsh, Ieuan A. Hughes, and Carlo L. Acerini, "Prenatal Androgen Exposure Alters Girls' Responses to Information Indicating Gender-Appropriate Behavior," *Philosophical Transactions of the Royal Society B: Biological Sciences* 371, no. 1688 (2016): 20150125; Heino F.L. Meyer-Bahlburg, Curtis Dolezal, Susan W. Baker, Anke A. Ehrhardt, and Maria I. New, "Gender Development in Women with Congenital Adrenal Hyperplasia as a Function of Disorder Severity," *Archives of Sexual Behavior* 35, no. 6 (2006): 667–84; Vickie Pasterski, Peter Hindmarsh, Mitchell Geffner, Charles Brook, Caroline Brain, and Melissa Hines, "Increased Aggression and Activity Level in 3- to 11-Year-Old Girls with Congenital Adrenal Hyperplasia (CAH)," *Hormones and Behavior* 52, no. 3 (2007): 368–74; Vickie Pasterski, Mitchell E. Geffner, Caroline Brain, Peter Hindmarsh, Charles Brook, and Melissa Hines, "Prenatal Hormones and Childhood Sex Segregation: Playmate and Play Style Preferences in Girls with Congenital Adrenal Hyperplasia,"

Hormones and Behavior 59, no. 4 (2011): 549–55; Debra Spencer, Vickie Paster-ski, Sharon A. S. Neufeld, Vivette Glover, Thomas G. O'Connor ... and Melissa Hines, "Prenatal Androgen Exposure and Children's Gender-Typed Behavior and Toy and Playmate Preferences," *Hormones and Behavior* 127 (2021): 104889, 1–9; Debra Spencer, Vickie Pasterski, Sharon Neufeld, Vivette Glover, Thomas G. O'Connor ... and Melissa Hines, "Prenatal Androgen Exposure and Children's Aggressive Behavior and Activity Level," *Hormones and Behavior* 96 (2017): 156–65.

97   CAH girls grow up to be women who: Adriene M. Beltz, Jane L. Swanson, and Sheri A. Berenbaum, "Gendered Occupational Interests: Prenatal Androgen Effects on Psychological Orientation to Things Versus People," *Hormones and Behavior* 60, no. 4 (2011): 313–17; and Sheri A. Berenbaum, "Beyond Pink and Blue: The Complexity of Early Androgen Effects on Gender Development," *Child Development Perspectives* 12, no. 1 (2018): 58–64.

97   Although medical practices are changing: For example, in a move toward reducing (or banning) medically unnecessary surgery on intersex children, Boston Children's Hospital recently announced that it will no longer perform genital surgery on children who are not old enough to understand and consent. Surgeons at the hospital will no longer reduce the size of a "too large" clitoris (that may resemble a penis), nor will they perform vaginoplasty on intersex children, in which something resembling a vagina is created. Shefali Luthra, "Boston Children's Hospital Will No Longer Perform Two Types of Intersex Surgery on Children," *USA Today*, October 22, 2020.

98   These social influences, in this view: Jordan-Young, "Hormones, Context, and 'Brain Gender.'"

98   girls are generally policed more leniently: Hugh Lytton and David M. Romney, "Parents' Differential Socialization of Boys and Girls: A Meta-Analysis," *Psychological Bulletin* 109, no. 2 (1991): 267–96.

98   parents report that they encourage: Celina C. C. Cohen-Bendahan, Cornelieke van de Beek, and Sheri A. Berenbaum, "Prenatal Sex Hormone Effects on Child and Adult Sex-Typed Behavior: Methods and Findings," *Neuroscience and Biobehavioral Reviews* 29, no. 2 (2005): 353–84. But see Wang I. Wong, Vickie Pasterski, Peter C. Hindmarsh, Mitchell E. Geffner, and Melissa Hines, "Are There Parental Socialization Effects on the Sex-Typed Behavior of Individuals with Congenital Adrenal Hyperplasia?," *Archives of Sexual Behavior* 42, no. 3 (2013): 381–91, for one (outlier) study that found parents give CAH girls more encouragement for play with boy-typical toys than unaffected daughters. However, the authors conclude that prenatal androgen exposure also contributes to sex-typical toy play.

98   what most kids seem to want: Kay Bussey and Albert Bandura, "Influence of Gender Constancy and Social Power on Sex-Linked Modeling," *Journal of Personality and Social Psychology* 47, no. 6 (1984): 1292–302.

98   Measurements of T in a pregnant mother's blood: Melissa Hines, "Prenatal Testosterone and Gender-Related Behaviour," *European Journal of Endocrinology* 155, suppl. 1 (2006): S115–S121.

99   levels of T on the high end of normal during a mom's pregnancy: Melissa
      Hines, "Prenatal Endocrine Influences on Sexual Orientation and on Sex-
      ually Differentiated Childhood Behavior," *Frontiers in Neuroendocrinology*
      32, no. 2 (2011): 170–82; Melissa Hines, "Human Gender Development,"
      Neuroscience and Biobehavioral Reviews 118 (2020): 89–96; but see Rebecca
      Christine Knickmeyer, Sally Wheelwright, Kevin Taylor, Peter Raggatt, Ger-
      ald Hackett, and Simon Baron-Cohen, "Gender-Typed Play and Amniotic
      Testosterone," *Developmental Psychology* 41, no. 3 (2005): 517–58, for null
      findings.

100   T makes boy brains: Sex differences in levels of sex hormones are not the only
      aspects of our physical selves that could directly affect behavior (as opposed
      to indirect effects, like genitalia or muscle differences). The sexes also differ in
      the number of X chromosomes, in addition to the genes on the Y chromosome
      that females do not possess. Some recent evidence suggests that differences
      in the dosages of X chromosome genes, along with the expression of male-
      specific Y-chromosome genes, may exert sex-specific effects on tissue develop-
      ment and function, including brain development and function during critical
      periods. These are two reviews that support the claim broadly: Daniel M. Snell
      and James M. A. Turner, "Sex Chromosome Effects on Male–Female Differ-
      ences in Mammals," *Current Biology* 28, no. 22 (2018): R1313–R24; Arthur P.
      Arnold, "Sexual Differentiation of Brain and Other Tissues: Five Questions for
      the Next 50 Years," *Hormones and Behavior* 120 (2020): 104691.

      For Y-chromosome gene expression in nonreproductive tissues, see
      Alexander K. Godfrey, Sahin Naqvi, Lukáš Chmátal, Joel M. Chick, Richard
      N. Mitchell, Steven P. Gygi, Helen Skaletsky, and David C. Page, "Quantita-
      tive Analysis of Y-Chromosome Gene Expression Across 36 Human Tissues,"
      Genome Research 30, no. 6 (2020): 860–73.

      For Y-chromosome gene expression in the brain, see Ivanka Savic,
      Louise Frisen, Amirhossein Manzouri, Anna Nordenstrom, and Angelica
      Lindén Hirschberg, "Role of Testosterone and Y Chromosome Genes for
      the Masculinization of the Human Brain," *Human Brain Mapping* 38, no. 4
      (2017): 1801–14.

      For sex differences in X-chromosome gene expression and its role in
      sex differences in disease, see Haiko Schurz, Muneeb Salie, Gerard Tromp,
      Eileen G. Hoal, Craig J. Kinnear, and Marlo Möller, "The X Chromosome
      and Sex-Specific Effects in Infectious Disease Susceptibility," *Human Genom-
      ics* 13, no. 1 (2019): 1–12.

CHAPTER 5

101   "one question is clear": Claire Watson, "Semenya Humiliated: Athletics
      Chief," Reuters, August 20, 2009, https://af.reuters.com/article/idAFJOE
      57J0NP20090820.

102   "just look at her": Christopher Clarey and Gina Kolata, "Gold Awarded amid
      Dispute over Runner's Sex," *New York Times*, August 20, 2009.

102   The gold was then awarded to the athlete next in line: Karolos Grahmann, "Savinova Stripped of London Games 800m Gold for Doping," Reuters, February 10, 2017.

102   "Could This Women's World Champ Be a Man?": William Lee Adams, "Could This Women's World Champ Be a Man?," *Time*, August 21, 2009.

102   The colorful spread inside the magazine: "Makeover for SA Gender-Row Runner," BBC News, September 8, 2009, http://news.bbc.co.uk/2/hi/8243553.stm; and Tracy Clark-Flory, "Sex Test Runner Gets a Girly Makeover," *Salon*, September 8, 2009, https://www.salon.com/2009/09/08/runner_makeover/.

102   "We have to see if she has an advantage": "Caster Semenya: Anatomy of Her Case," *Telegraph* (UK), July 6, 2010.

103   the IAAF introduced new regulations: Rick Maese, "Court Rules Olympic Runner Caster Semenya Must Use Hormone-Suppressing Drugs to Compete," *Washington Post*, May 1, 2019.

103   were meant to target her in particular: There is no evidence that the regulations were meant to target Semenya in particular. The IAAF at the time was evaluating the cases of other similar intersex athletes.

103   "biology has to trump gender identity": "IAAF Response to Swiss Federal Tribunal's Decision," World Athletics, Monaco, press release, July 31, 2019, https://www.worldathletics.org/news/press-release/swiss-federal-tribunal-decision.

103   she has refused to lower her T levels: Jacob Bogage, "Caster Semenya Blocked from Defending 800 Title at Worlds After Swiss Court Reverses Ruling," *Washington Post*, July 30, 2019.

103   Your body can't tell the difference: David J. Handelsman, Angelica L. Hirschberg, and Stephane Bermon, "Circulating Testosterone as the Hormonal Basis of Sex Differences in Athletic Performance," *Endocrine Reviews* 39, no. 5 (2018): 803–29.

103   she'd have no chance against the best men: "'But Seriously,' Tennis Great John McEnroe Says He's Seeking 'Inner Peace,'" *Weekend Edition Sunday*, National Public Radio, June 25, 2017, https://www.npr.org/2017/06/25/534149646/but-seriously-tennis-great-john-mcenroe-says-hes-seeking-inner-peace.

104   clearly irritated Williams: Cindy Boren, "Serena Williams vs. John McEnroe: It's Game, Set, Match Serena with a Nude Vanity Fair Cover to Boot," *Washington Post*, June 27, 2017.

104   "I only want to play girls": Evan Hilbert, "Serena Williams on Playing Andy Murray: 'I'd Lose 6–0, 6–0,'" CBS Sports, August 23, 2013.

105   The difference in throwing is vast: See David J. Epstein, *The Sports Gene: Inside the Science of Extraordinary Athletic Performance* (New York: Current, 2014), 56–74, for a comprehensive description of the development of sex differences in throwing and in athletic ability more generally.

105   "Nearly every boy by age 15 throws better": Jerry Thomas quoted in Tamar Haspel, "Throw Like a Girl? With Some Practice, You Can Do Better," *Washington Post*, September 10, 2012.

105   Long-distance swimming: Øyvind Sandbakk, Guro Strøm Solli, and Hans-Christer Holmberg, "Sex Differences in World-Record Performance: The

Influence of Sport Discipline and Competition Duration," *International Journal of Sports Physiology and Performance* 13, no. 1 (2018): 2–8; and Beat Knechtle et al., "Women Outperform Men in Ultra-Distance Swimming: The Manhattan Island Marathon Swim from 1983 to 2013," *International Journal of Sports Physiology and Performance* 9, no. 6 (2014): 913–24.

105 Sex differences in testosterone levels and sports performance: adapted from David J. Handelsman, "Sex Differences in Athletic Performance Emerge Coinciding with the Onset of Male Puberty," *Clinical Endocrinology* 87, no. 1 (2017): 68–72.

106 Women's world records are consistently about 10 percent: Sandbakk, Solli, and Holmberg, "Sex Differences in World-Record Performance."

106 In 2019, about *twenty-five hundred* men: "Season Top Lists: 100 Meters Men, 100 Meters Women," World Athletics (2019), accessed August 15, 2020, https://www .worldathletics.org/records/toplists/sprints/100-metres/outdoor/men/senior /2019?regionType=world&timing=electronic&windReading=regular&page =23&bestResultsOnly=true. About 8,100 men and 5,470 women (in the Senior, which is open age, and under twenty categories) competed in the IAAF 100-meter track event worldwide in 2019. Shelly-Ann Fraser-Pryce from Jamaica set the 2019 record with a time of 10.71 seconds, and Cristian Coleman set the record with a time of 9.76 seconds. About 2,100 senior men and 500 under twenty men ran the distance faster than 10.71 seconds, beating Fraser-Pryce's time.

107 "It's misleading to take the current gap": Fred Dreier, "Q&A: Dr. Rachel McKinnon, Masters Track Champion and Transgender Athlete," VeloNews, October 15, 2018, https://www.velonews.com/news/qa-dr-rachel-mckinnon-masters -track-champion-and-transgender-athlete/.

107 the gap hasn't budged: Mindy Millard-Stafford, Ann E. Swanson, and Matthew T. Wittbrodt, "Nature Versus Nurture: Have Performance Gaps Between Men and Women Reached an Asymptote?," *International Journal of Sports Physiology and Performance* 13, no. 4 (2018): 530–35; Valérie Thibault, Marion Guillaume, Geoffroy Berthelot et al., "Women and Men in Sport Performance: The Gender Gap Has Not Evolved Since 1983," *Journal of Sports Science and Medicine* 9, no. 2 (2010): 214–23; but see Laura Capranica, Maria F. Piacentini, Shona Halson et al., "The Gender Gap in Sport Performance: Equity Influences Equality," *International Journal of Sports Physiology and Performance* 8, no. 1 (2013): 99–103, for the argument that until women and men have equal opportunities and compensation in sports, we can't know the limits to women's sports performance.

107 From 1972 to 1980, the sex differences in Olympic trial times went from: Millard-Stafford et al., "Nature Versus Nurture," *International Journal of Sports Physiology and Performance* 13, no. 4 (2018): 530–535. For sex differences in Olympic performance over time see Thibault et al., "Women and Men in Sports Performance," 214.

108 "perhaps they would up their performance": Beth Jones quoted in Sean Ingle, "Why Calls for Athletes to Compete as a Homogenised Group Should Be Resisted," *Guardian*, December 10, 2017.

108 *The more testosterone, the better the athlete* is myth number five: Rebecca M. Jordan-Young and Katrina Karkazis, "Five Myths About Testosterone," *Washington Post*, October 25, 2019.

109 "cast doubt on the thesis that T is a master key to": Jordan-Young and Karkazis, "Five Myths About Testosterone."

109 In endurance sports: Anthony C. Hackney and Amy R. Lane, "Low Testosterone in Male Endurance-Trained Distance Runners: Impact of Years in Training," *Hormones* 17, no. 1 (2018): 137–39; Javier Alves, Víctor Toro, Gema Barrientos et al., "Hormonal Changes in High-Level Aerobic Male Athletes During a Sports Season," *International Journal of Environmental Research and Public Health* 17, no. 16 (2020): 5833; and see S. Bermon and P. Y. Garnier, "Serum Androgen Levels and Their Relation to Performance in Track and Field: Mass Spectrometry Results from 2127 Observations in Male and Female Elite Athletes," *British Journal of Sports Medicine* 51, no. 17 (2017): 1309–14, for results in women, which show more of a consistently positive relationship between T levels and performance in most track and field events.

110 Since that bound T can't get inside the cell: The "classical" action of steroid hormones, that is, the one that is best documented, is via direct action on gene transcription, aka "genomic" action (as I've described in the main text). But steroids can act through a different, faster pathway, although this mode of action is not well understood. This is called "nongenomic" action, in which steroid hormones interact with membrane-bound protein receptors rather than receptors inside the cell, in the nucleus or cytoplasm, and then go on to affect gene transcription. This is a fascinating area of research that opens up possible mechanisms for rapid actions of steroids that would affect body or behavior in a shorter time frame than is possible with genomic action. It appears likely that rapid, nongenomic actions of steroids also shape future genomic actions. For a review of nongenomic actions of steroids, including androgens, see Sandi R. Wilkenfeld, Chenchu Lin, and Daniel E. Frigo, "Communication Between Genomic and Non-Genomic Signaling Events Coordinate Steroid Hormone Actions," *Steroids* 133 (2018): 2–7.

110 It depends on the type of test: Mathis Grossmann, "Utility and Limitations in Measuring Testosterone," in *Testosterone: From Basic to Clinical Aspects*, ed. Alexandre Hohl (Cham, Switzerland: Springer International, 2017), 97–107.

111 "cross reactivity" with other steroids: Another reason that RIA tends to inflate women's T levels is that microscopic blood particles often contaminate the saliva samples, which also inflate the T results. See Katie T. Kivlighan, Douglas A. Granger, Eve B. Schwartz, Vincent Nelson, Mary Curran, and Elizabeth A. Shirtcliff, "Quantifying Blood Leakage into the Oral Mucosa and Its Effects on the Measurement of Cortisol, Dehydroepiandrosterone, and Testosterone in Saliva," *Hormones and Behavior* 46, no. 1 (2004): 39–46. For more evidence on the errors and inconsistencies in steroid measurement created by cross-reactivity in RIA, see Frank Z. Stanczyk, Michael M. Cho, David B. Endres, John

L. Morrison, Stan Patel, and Richard J. Paulson, "Limitations of Direct Estradiol and Testosterone Immunoassay Kits," *Steroids* 68, no. 14 (2003): 1173–78.

111 "This tendency to inflate very low concentrations of testosterone": Keith M. Welker, Bethany Lassetter, Cassandra Brandes et al., "A Comparison of Salivary Testosterone Measurement Using Immunoassays and Tandem Mass Spectrometry," *Psychoneuroendocrinology* 71 (2016): 180–88.

111 "Guessing would be more accurate": David A. Herold and Robert L. Fitzgerald, "Immunoassays for Testosterone in Women: Better Than a Guess?," *Clinical Chemistry* 49, no. 8 (2003): 1250–51.

111 we should be cautious about claims: Valérie Moal, Elisabeth Mathieu, Pascal Reyner et al., "Low Serum Testosterone Assayed by Liquid Chromatography-Tandem Mass Spectrometry. Comparison with Five Immunoassay Techniques," *Clinica Chimica Acta* 386, no. 1 (2007): 12–19.

111 "in reality, testosterone levels *overlap considerably*": Sari M. van Anders, Zach C. Schudson, Emma C. Abed et al., "Biological Sex, Gender, and Public Policy," *Policy Insights from the Behavioral and Brain Sciences* 4, no. 2 (2017): 194–201.

111 She is not the only scholar to say so: For similar claims about T overlap between the sexes, see Katrina Karkazis and Rebecca Jordan-Young, "Debating a Testosterone 'Sex Gap,'" *Science* 348, no. 6237 (2015): 858–60; and Cara Tannenbaum and Sheree Bekker, "Sex, Gender, and Sports," editorial, *BMJ* 364 (2019): 1120.

111 "Usually, [the binary] is there for political reasons": Allison Whitten, "Untangling Gender and Sex in Humans," *Discover*, July 23, 2020.

112 The most recent, comprehensive, and rigorous study: Handelsman, Hirschberg, and Bermon, "Circulating Testosterone as the Hormonal Basis of Sex Differences in Athletic Performance."

112 this method is the most accurate: Even when using mass spectrometry for steroid analysis, blood samples allow a more accurate measurement of serum T than do saliva samples. See Tom Fiers, Joris Delanghe, Guy T'Sjoen, Eva Van Caenegem, Katrien Wierckx, and Jean-Marc Kaufman, "A Critical Evaluation of Salivary Testosterone as a Method for the Assessment of Serum Testosterone," *Steroids* 86 (2014): 5–9; and B. G. Keevil, P. MacDonald, W. Macdowall, D. M. Lee, F. C. W. Wu, and NATSAL Team. "Salivary Testosterone Measurement by Liquid Chromatography Tandem Mass Spectrometry in Adult Males and Females," *Annals of Clinical Biochemistry* 51, no. 3 (2014): 368–78.

113 "wide and complete separation between men and women": Handelsman, Hirschberg, and Bermon, "Circulating Testosterone as the Hormonal Basis of Sex Differences in Athletic Performance," 806.

113 like two mountains separated by a vast plain: Ilustration adapted with permission from Doriane L. Coleman, "Sex in Sport," *Law and Contemporary Problems* 80 (2017): 63–126, based on original data from Richard V. Clark, Jeffrey A. Wald, Ronald S. Swerdloff, Christina Wang, Frederick C. W. Wu, Larry D. Bowers, and Alvin M. Matsumoto, "Large Divergence in Testosterone

Concentrations Between Men and Women: Frame of Reference for Elite Athletes in Sex-Specific Competition in Sports, a Narrative Review," *Clinical Endocrinology* 90, no. 1 (2019): 15–22 .

114   Polycystic ovarian syndrome (PCOS): Valentina Rodriguez Paris and Michael J. Bertoldo, "The Mechanism of Androgen Actions in PCOS Etiology," *Medical Sciences* (Basel, Switzerland) 7, no. 9 (2019): 1–12.

114   some men don't have testicles at all: For various nonmedical reasons, some men are "voluntary" eunuchs, meaning, they desire to be castrated. For a review, see Thomas W. Johnson, Michelle A. Brett, Lesley F. Roberts, and Richard J. Wassersug, "Eunuchs in Contemporary Society: Characterizing Men Who Are Voluntarily Castrated (Part I)," *Journal of Sexual Medicine* 4, no. 4 (2007): 930–45.

115   5-ARD is common: "IAAF Publishes Briefing Notes and Q&A on Female Eligibility Regulations," World Athletics, press release, May 7, 2019, https://www.worldathletics.org/news/press-release/questions-answers-iaaf-female-eligibility-reg.

115   The graphic above illustrates the T ranges of people: Illustration adapted with permission from Coleman, "Sex in Sport," based on original data from Clark et al., "Large Divergence in Testosterone Concentrations Between Men and Women."

116   T levels peaked around age twenty . . . and then slowly declined: Shalender Bhasin, Michael Pencina, Guneet Kaur Jasuju et al., "Reference Ranges for Testosterone in Men Generated Using Liquid Chromatography Tandem Mass Spectrometry in a Community-Based Sample of Healthy Nonobese Young Men in the Framingham Heart Study and Applied to Three Geographically Distinct Cohorts," *Journal of Clinical Endocrinology and Metabolism* 96, no. 8 (2011): 2430–39; and S. Mitchell Harman, E. Jeffrey Metter, Jordan D. Tobin, Jay Pearson, and Marc R. Blackman, "Longitudinal Effects of Aging on Serum Total and Free Testosterone Levels in Healthy Men," *Journal of Clinical Endocrinology and Metabolism* 86, no. 2 (2001): 724–31. Age-related drops in T level are not always found, particularly in non-Western, less urban societies, where male T levels tend to be lower. See Peter T. Ellison and Catherine Panter-Brick, "Salivary Testosterone Levels Among Tamang and Kami Males of Central Nepal," *Human Biology* 68, no. 6 (1996): 955–65; and Peter T. Ellison, Richard G. Bribiescas, Gillian R. Bentley et al., "Population Variation in Age-Related Decline in Male Salivary Testosterone," *Human Reproduction* 17, no. 12 (2002): 3251–53. For an overview of T and its actions across the life span in men, see Richard G. Bribiescas, "Reproductive Ecology and Life History of the Human Male," *American Journal of Physical Anthropology* 116, no. S33 (2001): 148–76.

118   Although Sofia's growth spurt started before Sam's: Ana Paula Abreu and Ursula B. Kaiser, "Pubertal Development and Regulation," *Lancet: Diabetes and Endocrinology* 4, no. 3 (2016): 254–64.

119   Higher T in Sam also enlarges: Karen L. Herbst and Shalender Bhasin, "Testosterone Action on Skeletal Muscle," *Current Opinion in Clinical Nutrition*

& *Metabolic Care* 7, no. 3 (2004): 271–77; and James G. MacKrell, Benjamin C. Yaden, Heather Bullock et al., "Molecular Targets of Androgen Signaling That Characterize Skeletal Muscle Recovery and Regeneration," *Nuclear Receptor Signaling* 13, no. 1 (2015): 1–19.

119   Sofia is not going to win a pull-up contest: Phillip Bishop, Kirk Cureton, and Mitchell Collins, "Sex Difference in Muscular Strength in Equally-Trained Men and Women," *Ergonomics* 30, no. 4 (1987): 675–87; and J. C. Wells, "Sexual Dimorphism of Body Composition," *Best Practice and Research in Clinical Endocrinology and Metabolism* 21, no. 3 (2007): 415–30.

120   Bones have receptors for both estrogen and testosterone: In childhood, Sofia's and Sam's bones grew at about the same relatively slow rate (compared to puberty), primarily under the influence of growth hormone and IGF-1. Those tissue-building hormones take the lead in transforming babies into little kids. In puberty, the rapid rise of sex hormones augments the effects of growth hormone and IGF-1, so that bones grow faster.

120   When estrogen was given instead: Daniela Merlotti, Luigi Gennari, Stolakis Konstantinos, and Nuti Ranuccio, "Aromatase Activity and Bone Loss in Men," *Journal of Osteoporosis* 2011 (2011), article 230671.

120   Girls may briefly overtake boys in height: Christine Wohlfahrt-Veje, Annette Mouritsen, Casper P. Hagen et al., "Pubertal Onset in Boys and Girls Is Influenced by Pubertal Timing of Both Parents," *Journal of Clinical Endocrinology and Metabolism* 101, no. 7 (2016): 2667–74.

121   This effect of T is not permanent: Androgens positively affect hemoglobin levels. For effects in men, see Shalender Bhasin, Linda Woodhouse, Richard Casaburi et al., "Testosterone Dose-Response Relationships in Healthy Young Men," *American Journal of Physiology-Endocrinology and Metabolism* 281, no. 6 (2001): 1172–81. For effects of large testosterone changes in transgender people taking or blocking male levels of testosterone as part of a gender transition, see Denise Chew, Jemma Anderson, Katrina Williams, Tamara May, and Kenneth Pang, "Hormonal Treatment in Young People with Gender Dysphoria: A Systematic Review," *Pediatrics* 141, no. 4 (2018): e20173742.

122   "*T alone didn't do much*": Rebecca M. Jordan-Young and Katrina Karkazis, *Testosterone: An Unauthorized Biography* (Cambridge, MA: Harvard University Press, 2019), 289 (my emphasis).

124   these findings have been replicated: Replication of Bhasin's studies: Joel S. Finkelstein, Hang Lee, Sherri-Ann Burnett-Bowie et al., "Gonadal Steroids and Body Composition, Strength, and Sexual Function in Men," *New England Journal of Medicine* 369, no. 11 (2013): 1011–22. Also see Stefan M. Pasiakos, Claire E. Berryman, J. Philip Karl et al., "Effects of Testosterone Supplementation on Body Composition and Lower-Body Muscle Function During Severe Exercise- and Diet-Induced Energy Deficit: A Proof-of-Concept, Single Centre, Randomised, Double-Blind, Controlled Trial," *EBioMedicine* 46 (2019): 411–22.

125   people with these conditions are overrepresented: S. Bermon, P. Y. Garnier, A. L. Hirschberg et al., "Serum Androgen Levels in Elite Female Athletes," *Journal of Clinical Endocrinology and Metabolism* 99, no. 11 (2014): 4328–35.

125   The upper limit of T for women with PCOS: Handelsman, Hirschberg, and
       Bermon, "Circulating Testosterone as the Hormonal Basis of Sex Differences
       in Athletic Performance."

125   37 percent had PCOS: Magnus Hagmar, Bo Berglund, Kerstin Brismar, and
       Angelica L. Hirschberg, "Hyperandrogenism May Explain Reproductive
       Dysfunction in Olympic Athletes," *Medicine and Science in Sports and Exer-
       cise* 41, no. 6 (2009): 1241–48.

126   "Female track athletes with naturally": Doug Mills, "Caster Semenya
       Loses Case to Compete as a Woman in All Races," *New York Times*, May
       1, 2019. Athletes who fall under the IAAF DSD regulations must have
       XY sex chromosomes, testicles, and testosterone levels in the typical male
       range. See "IAAF Publishes Briefing Notes and Q&A on Female Eligibility
       Regulations."

127   these tests have been inaccurate: For a history of sex testing in sports (along
       with a proposal for an "athletic gender"), see Joanna Harper, "Athletic Gen-
       der," *Law and Contemporary Problems* 80, no. 4 (2017): 98–110.

127   "'They say I can't run and I have to go home'": Deborah Larned, "The Fem-
       ininity Test: A Woman's First Olympic Hurdle," *Womensports* 3 (1976):
       8, as cited in V. Heggie, "Testing Sex and Gender in Sports; Reinventing,
       Reimagining and Reconstructing Histories," *Endeavour* 34, no. 4 (December
       2010): 157–63.

128   effects of reduced T may be different: Anna Wiik, Tommy R. Lundberg,
       Eric Rullman et al., "Muscle Strength, Size and Composition Following
       12 Months of Gender-Affirming Treatment in Transgender Individu-
       als," *Journal of Clinical Endocrinology and Metabolism* 105, no. 3 (2019):
       e805–e813.

128   "lack a sound scientific basis": Court of Arbitration for Sport, Executive
       Summary, retrieved August 15, 2020, https://www.tas-cas.org/fileadmin
       /user_upload/CAS_Executive_Summary__5794_.pdf, 2.

129   "The Panel pays tribute to Ms. Semenya's grace and fortitude": Court of Arbi-
       tration for Sport, Executive Summary, 6.

CHAPTER 6

130   deer population about 1,000: "Red Deer," Isle of Rum website, Isle of Rum
       Community Trust, updated January 2020, http://www.isleofrum.com
       /wildlifedeer.php.

132   sweet spot is between seven and ten years: T. H. Clutton-Brock, S. D. Albon,
       R. M. Gibson, and F. E. Guinness, "The Logical Stag: Adaptive Aspects of
       Fighting in Red Deer (*Cervus elaphus* L.)," *Animal Behaviour* 27 (1979):
       211–25.

132   each season only a few succeed: Red deer have some characteristics and hab-
       its that make it relatively easy for stags to defend a group of hinds. They live
       on land rather than in the air or water, for one thing. Two-dimensional space
       makes it easier for an animal to defend a resource than three-dimensional

space—that's why harems are very rare in fish and birds. Think of all the ways that SF fish or birds could weasel their way into another's group! And red deer females congregate in groups. Defending a group is difficult if you first have to get the group together. For a discussion of how two- versus three-dimensional space impacts sexual selection, see David Puts, "Beauty and the Beast: Mechanisms of Sexual Selection in Humans," *Evolution and Human Behavior* 31 (May 1, 2010): 157–75.

133   it can pay off with an actual mating: Clutton-Brock et al., "The Logical Stag."

134   red deer stags don't make a habit of recklessly jumping into a physical conflict: For a description of fighting between stags, including assessing threat, stages of escalation, and costs of fighting, see the fascinating book on Rum research by Tim H. Clutton-Brock, Fiona E. Guinness, and Steve D. Albon, *Red Deer: Behavior and Ecology of Two Sexes* (Chicago: University of Chicago Press, 1982), 128–39.

134   qualities that are needed to be victorious in battle: Importance of vocalizations: David Reby, Karen McComb, Bruno Cargnelutti et al., "Red Deer Stags Use Formants as Assessment Cues During Intrasexual Agonistic Interactions," *Proceedings of the Royal Society B: Biological Sciences* 272, no. 1566 (2005): 941–47.

134   strength and body size: Clutton-Brock et al., "The Logical Stag."

134   "behavioral factors apparently affect an individual's chance of winning": Clutton-Brock et al., "The Logical Stag," 218–19.

135   If one goes down: Rarely stags' antlers get stuck together, and the stags can die from starvation. Rebecca Nagy, "Fighting Bucks Get Their Horns Stuck Together," *Roaring Earth*, n.d., https://roaring.earth/fighting-bucks-get-stuck/.

137   being top hind can pay off: Relative to males, reproductively successful hinds will accrue smaller benefits over a longer reproductive career.

137   A successful hind might bear a calf every year: R. M. Gibson and F. E. Guinness, "Differential Reproduction Among Red Deer (*Cervus elaphus*) Stags on Rhum," *Journal of Animal Ecology* 49, no. 1 (1980): 199–208; and Roger Lewin, "Red Deer Data Illuminate Sexual Selection," *Science* 218, no. 4578 (1982): 1206–8. Not all the hinds in a harem get pregnant during one breeding season, and harem membership isn't stable throughout a season.

137   Last year he sired about fifteen calves: Wisdom 11's paternity data from 2019 are from personal communication with Rum's research director, Josephine Pemberton.

137   fuel different reproductive strategies: Dominance is not always associated with increased reproductive success in either sex, and it is one of many strategies. See Marlene Zuk, *Sexual Selections: What We Can and Can't Learn About Sex from Animals* (Berkeley: University of California Press, 2002), 124–28.

137   during the rut, when he spends only about 5 percent: Clutton-Brock, Guinness, and Albon, *Red Deer*, 121–22.

138   They grow in size and triple in weight: For testes size changes across the seasons, see A. F. Malo, E. R. S. Roldan, J. J. Garde et al., "What Does Testosterone

Do for Red Deer Males?," *Proceedings of the Royal Society B: Biological Sciences* 276, no. 1658 (2008): 971–80. For changes in T levels and testes weight, see G. A. Lincoln, "The Seasonal Reproductive Changes in the Red Deer Stag (*Cervus elaphus*)," *Journal of Zoology* 163, no. 1 (1971): 105–23; and G. A. Lincoln and R. N. B. Kay, "Effects of Season on the Secretion of LH and Testosterone in Intact and Castrated Red Deer Stags (*Cervus elaphus*)," *Journal of Reproduction and Fertility* 55, no. 1 (1979): 75–80.

138   A longer vocal tract produces a deeper voice: Benjamin D. Charlton, David Reby, and Karen McComb, "Female Red Deer Prefer the Roars of Larger Males," *Biology Letters* 3, no. 4 (2007): 382–85.

139   The velvet provides a blood supply to the growing bones: S. Gomez, A. J. Garcia, S. Luna et al., "Labeling Studies on Cortical Bone Formation in the Antlers of Red Deer (*Cervus elaphus*)," *Bone* 52, no. 1 (2013): 506–15.

139   As in humans, high T increases the calcification of bone: Malo et al., "What Does Testosterone Do for Red Deer Males?"; and Gomez et al., "Labeling Studies on Cortical Bone Formation."

140   T also encourages the growth of a shaggy mane: E. Gaspar-López, T. Landete-Castillejos, J. A. Estevez et al., "Seasonal Variations in Red Deer (*Cervus elaphus*) Hematology Related to Antler Growth and Biometrics Measurements," *Journal of Experimental Zoology Part A: Ecological Genetics and Physiology* 315, no. 4 (2011): 242–49; and David Granville Thomas, "The Hormonal Control of Hair Growth in the Red Deer (*Cervus elaphus*)" (PhD diss., University College London, 1997).

140   More oxygen to working muscles: Malo et al., "What Does Testosterone Do for Red Deer Males?"

141   sugar hypothesis: Mark L. Wolraich, David B. Wilson, and J. Wade White, "The Effect of Sugar on Behavior or Cognition in Children: A Meta-Analysis," *JAMA* 274, no. 20 (1995): 1617–21.

142   Such experiments were carried out: G. A. Lincoln, Fiona Guinness, and R. V. Short, "The Way in Which Testosterone Controls the Social and Sexual Behavior of the Red Deer Stag (*Cervus elaphus*)," *Hormones and Behavior* 3, no. 4 (1972): 375–96.

144   male and female forms: Sexual selection acts strongly on males in species in which male mate competition is intense, as in the red deer, but it also acts on female animals. For an overview, see T. H. Clutton-Brock and Elise Huchard, "Social Competition and Selection in Males and Females," *Philosophical Transactions of the Royal Society B: Biological Sciences* 368, no. 1631 (2013): 20130074.

144   These differences lead to a predictable, sex-based behavioral pattern: T. H. Clutton-Brock and G. A. Parker, "Potential Reproductive Rates and the Operation of Sexual Selection," *Quarterly Review of Biology* 67, no. 4 (1992): 437–56.

For a thorough overview of sexual selection, see David C. Geary, *Male, Female: The Evolution of Human Sex Differences,* 3rd ed. (Washington, DC: American Psychological Association, 2021), 67–140.

For one of the most important contributions to explanations for differences between the sexes (as an ultimate consequence of differences in parental investment), see Robert Trivers, "Parental Investment and Sexual Selection," in *Sexual Selection and the Descent of Man, 1871–1971*, ed. Bernard Campbell, 136–79 (New York: Aldine de Gruyter, 1972).

For how sex differences in parental investment shape mating strategies, see Donald Symons, *The Evolution of Human Sexuality* (New York: Oxford University Press, 1979), 23–25.

Not all female mammals host their offspring development internally; the monotremes, including the platypus and four types of echidnas, lay eggs. But like all mammals, they produce milk to feed their young. And they use a diverse array of mating strategies. For example, in animals like the red-necked phalarope, after the female has laid her eggs, she leaves dad to do the tending, and it's the females that compete intensely for mates. In general, the ecological context, including factors such as the ratio of reproductively viable males to females, heavily impacts the nature and extent of sex differences in mating strategies. See Clutton-Brock and Parker, "Potential Reproductive Rates and the Operation of Sexual Selection," for an overview.

144   "depends, not on a struggle for existence in relation to other organic beings": Charles Darwin, *On the Origin of Species by Means of Natural Selection, Or Preservation of Favoured Races in the Struggle for Life* (London: John Murray, 1859), 87–88.

145   Competition between one sex (as Darwin says, usually males): Females of many species have also evolved adaptations that allow them to compete for mates, sometimes using overt physical aggression, or via more passive strategies such as the selection of the "best" sperm once already in the reproductive tract (cryptic choice). Females may select males based on genetic quality, resource provision, or parental investment, among other factors. For an overview, see Kimberly A. Rosvall, "Intrasexual Competition in Females: Evidence for Sexual Selection?," *Behavioral Ecology* 22, no. 6 (2011): 1131–40.

145   As Darwin put it: Darwin, *On the Origin of Species*, 88. Again, since Darwin's time scientists have documented the active role that females take in not only mate choice but also mating competition.

145   "The sight of a feather in a peacock's tail": Charles Darwin to Asa Gray, April 3 [1860], available through the Darwin Correspondence Project, Letter no. 2743, University of Cambridge, https://www.darwinproject.ac.uk/letter/DCP-LETT-2743.xml.

145   Darwin explained the ornamentation of male birds: Charles Darwin, *The Descent of Man, and Selection in Relation to Sex*, 2 vols. (New York: D. Appleton, 1871), vol. 1, 422.

146   When a female actively chooses: For a brief history of the initial resistance and eventual appreciation for the role of female choice in sexual selection, see Zuk, *Sexual Selections*, 7–10.

146　high rates and intensity of aggression in female animals: See Zuk, *Sexual Selections*, 128–30, for examples of aggressive tactics in female animals.

146　The queen naked mole rat: Jeffrey A. French, Aaryn C. Mustoe, Jon Cavanaugh, and Andrew K. Birnie, "The Influence of Androgenic Steroid Hormones on Female Aggression in 'Atypical' Mammals," *Philosophical Transactions of the Royal Society B: Biological Sciences* 368, no. 1631 (2013): 1–10.

146　looks exactly like a penis: Stephen E. Glickman, Gerald R. Cunha, Christine M. Drea, Alan J. Conley, and Ned J. Place, "Mammalian Sexual Differentiation: Lessons from the Spotted Hyena," *Trends in Endocrinology and Metabolism* 17, no. 9 (2006): 349–56. Female spotted hyenas give birth, urinate, and receive penetration by the male's penis during sex through a single orifice, an opening at the tip of the penis-like clitoris. They are the only female mammals without a vaginal opening.

146　killing off the babies of her female rivals: T. H. Clutton-Brock, S. J. Hodge, G. Spong et al., "Intrasexual Competition and Sexual Selection in Cooperative Mammals," *Nature* 444, no. 7122 (2006): 1065–68.

149　Not too high, nor too low, but just right: Michael C. Moore, "Testosterone Control of Territorial Behavior: Tonic-Release Implants Fully Restore Seasonal and Short-Term Aggressive Responses in Free-Living Castrated Lizards," *General and Comparative Endocrinology* 70, no. 3 (1988): 450–59.

149　if you castrate lizards and remove T during the fall breeding season: Michael C. Moore and Catherine A. Marler, "Effects of Testosterone Manipulations on Nonbreeding Season Territorial Aggression in Free-Living Male Lizards, *Sceloporus jarrovi*," *General and Comparative Endocrinology* 65, no. 2 (1987): 225–32.

149　if you take one of those normal, territorial, T-in-the-low-to-middle-zone *summer* lizards: Michael C. Moore, "Elevated Testosterone Levels During Nonbreeding-Season Territoriality in a Fall-Breeding Lizard, *Sceloporus jarrovi*," *Journal of Comparative Physiology A* 158, no. 2 (1986): 159–63. For T changes across seasons, see Moore and Marler, "Effects of Testosterone Manipulations." For effects of T manipulations on territorial aggression in the summer, when territories are being established and aggression is not yet at its highest level, see Moore, "Testosterone Control of Territorial Behavior," 457.

149　The effects of T: Lizards, like many seasonal breeders, are heavily influenced by environmental factors, including other males, temperature, daylight exposure, etc.

153　Challenge Hypothesis: John C. Wingfield, Robert E. Hegner, Alfred M. Dufty, and Gregory F. Ball, "The 'Challenge Hypothesis': Theoretical Implications for Patterns of Testosterone Secretion, Mating Systems, and Breeding Strategies," *American Naturalist* 136, no. 6 (1990): 829–46.

153　Although Wingfield's initial formulation of his Challenge Hypothesis has been revised: John C. Wingfield, Marilyn Ramenofsky, Robert E. Hegner, and Gregory F. Ball, "Whither the Challenge Hypothesis?," *Hormones and Behavior* 123 (2020): 104588.

153   males must compete intensely: John C. Wingfield, Sharon E. Lynn, and Kiran K. Soma, "Avoiding the 'Costs' of Testosterone: Ecological Bases of Hormone-Behavior Interactions," *Brain, Behavior and Evolution* 57, no. 5 (2001): 239–51.

153   The ups *and* downs of T are adaptive: Wingfield, Lynn, and Soma, "Avoiding the 'Costs' of Testosterone."

153   "It is in the management of this system": Peter T. Ellison, *On Fertile Ground: A Natural History of Human Reproduction* (Cambridge, MA: Harvard University Press, 2009), 260.

CHAPTER 7

155   as far as he could tell, he was in better physical shape: Humans may have evolved a specialized mechanism that rapidly allows them to assess a man's ability to fight or hold resources. See Aaron Sell, Leda Cosmides, John Tooby, Daniel Sznycer, Christopher von Rueden, and Michael Gurven, "Human Adaptations for the Visual Assessment of Strength and Fighting Ability from the Body and Face," *Proceedings of the Royal Society B: Biological Sciences* 276, no. 1656 (2009): 575–84.

155   Fairless recounted how things quickly got out of hand in his 2018 book: Daemon Fairless, *Mad Blood Stirring: The Inner Lives of Violent Men* (Toronto: Random House Canada, 2018), 4–7.

156   "The concept of gender roles is not cast as a biological phenomenon": American Psychological Association, "Harmful Masculinity and Violence," *In the Public Interest* newsletter, September 2018, https://www.apa.org/pi/about/newsletter/2018/09/harmful-masculinity.

156   "If you believe that T says something meaningful": Matthew Gutmann, "Testosterone Is Widely, and Wildly, Misunderstood," *Psyche* newsletter, Aeon, March 10, 2020, https://aeon.co/ideas/testosterone-is-widely-and-sometimes-wildly-misunderstood.

158   "Why don't you rape them before you kill them": Peter Landesman, "A Woman's Work," *New York Times*, September 15, 2002.

158   report in their book *Female Aggression*: For an overview of rates and types of women's physical aggression against intimate partners, see Helen Gavin and Theresa Porter, *Female Aggression* (Hoboken, NJ: John Wiley and Sons, 2014), 64–68.

158   Another group of researchers: John Archer, "Sex Differences in Aggression Between Heterosexual Partners: A Meta-Analytic Review," *Psychological Bulletin* 126, no. 5 (2000): 651–80; Sherry L. Hamby, "Measuring Gender Differences in Partner Violence: Implications from Research on Other Forms of Violent and Socially Undesirable Behavior," *Sex Roles* 52, no. 11–12 (2005): 725–42; and Murray A. Straus, "Dominance and Symmetry in Partner Violence by Male and Female University Students in 32 Nations," *Children and Youth Services Review* 30, no. 3 (2008): 252–75.

159   males engage in lower rates of behaviors related to empathy: Leonardo Christov-Moore, Elizabeth A. Simpson, Gino Coudé, Kristina Grigaityte,

Marco Iacobini, and Pier Francesco Ferrari, "Empathy: Gender Effects in Brain and Behavior," *Neuroscience and Biobehavioral Reviews* 46, pt. 4 (2014): 604–27.

159   Men ... dominate in the most extreme form of intimate-partner violence: Margo Wilson and Martin Daly, "Lethal and Nonlethal Violence Against Wives and the Evolutionary Psychology of Male Sexual Proprietariness," in *Rethinking Violence Against Women*, ed. Russell Dobash (Thousand Oaks, CA: Sage, 1998), 224; Chelsea M. Spencer and Sandra M. Stith, "Risk Factors for Male Perpetration and Female Victimization of Intimate Partner Homicide: A Meta-Analysis," *Trauma, Violence, and Abuse* 21, no. 3 (2020): 527–40.

160   Women's motivations are more often self-defense: Why women kill intimate partners: Wilson and Daly, "Lethal and Nonlethal Violence Against Wives"; Nancy C. Jurik and Russ Winn, "Gender and Homicide: A Comparison of Men and Women Who Kill," *Violence and Victims* 5, no. 4 (1990): 227–42; Kenneth Polk and David Ranson, "The Role of Gender in Intimate Homicide," *Australian and New Zealand Journal of Criminology* 24, no. 1 (1991): 15–24; Lisa D. Brush, "Violent Acts and Injurious Outcomes in Married Couples: Methodological Issues in the National Survey of Families and Households," *Gender and Society* 4, no. 1 (1990): 56–67; Shilan Caman, Katarina Howner, Marianne Kristiansson, and Joakim Sturup, "Differentiating Male and Female Intimate Partner Homicide Perpetrators: A Study of Social, Criminological and Clinical Factors," *International Journal of Forensic Mental Health* 15, no. 1 (2016): 26–34.

160   "Indirect aggression": John Archer and Sarah M. Coyne, "An Integrated Review of Indirect, Relational, and Social Aggression," *Personality and Social Psychology Review* 9, no. 3 (2005): 212–30. For an evolutionary perspective on how language allows the social transmission of information about individuals (gossip) that affects reputations, see Richard Wrangham, *The Goodness Paradox: The Strange Relationship Between Virtue and Violence in Human Evolution* (New York: Pantheon, 2019), 135–36.

For an overview of types and possible neuroendocrine mediators of women's aggression and competition, see Thomas F. Denson, Siobhan M. O'Dean, Khandis R. Blake, and Joanne R. Beames, "Aggression in Women: Behavior, Brain and Hormones," *Frontiers in Behavioral Neuroscience* 12 (2018): 81.

160   girls and women seem to have an affinity for it: For evidence that females prefer indirect to direct (and physical) aggression, see: Joyce F. Benenson, Henry Markovits, Brittany Hultgren, Tuyet Nguyen, Grace Bullock, and Richard Wrangham, "Social Exclusion: More Important to Human Females Than Males," *PLoS One* 8, no. 2 (2013): e55851; Joyce F. Benenson, Henry Markovits, Melissa Emery Thompson, and Richard W. Wrangham, "Under Threat of Social Exclusion, Females Exclude More Than Males," *Psychological Science* 22, no. 4 (2011): 538–44; and Steven Arnocky and Tracy Vaillancourt, "Sexual Competition Among Women: A Review of the Theory and Supporting Evidence," in *The Oxford Handbook of Women and Competition*, ed. Maryanne L. Fisher, 25–39 (New York: Oxford University Press, 2017).

161 Because female aggression generally serves different purposes: On hormones and female aggression, see Kristina O. Smiley, Sharon R. Ladyman, Papillon Gustafson, David R. Grattan, and Rosemary S. E. Brown, "Neuroendocrinology and Adaptive Physiology of Maternal Care," *Current Topics in Behavioral Neuroscience* 43 (2019): 161–210. The expression of female aggression, like that of males, is contingent on physiological and environmental circumstances, which appear to have a greater effect on the neuroendocrine modulation of aggression than does sex per se. For a review, see Natalia Duque-Wilckens and Brian C. Trainor, "Behavioral Neuroendocrinology of Female Aggression," in *Oxford Research Encyclopedias*: Neuroscience, 1–55 (New York: Oxford University Press, 2017).

161 women are just as disposed to anger as men: John Archer, "Sex Differences in Aggression in Real-World Settings: A Meta-Analytic Review," *Review of General Psychology* 8, no. 4 (2004): 291–322.

161 That's reactive aggression: Richard W. Wrangham, "Two Types of Aggression in Human Evolution," *Proceedings of the National Academy of Sciences* 115, no. 2 (2018): 245–53; and A. Siegel and J. Victoroff, "Understanding Human Aggression: New Insights from Neuroscience," *International Journal of Law and Psychiatry* 32, no. 4 (2009): 209–15.

162 Neither sex has a monopoly on either kind: Wrangham, "Two Types of Aggression in Human Evolution." War is an exception; it is planned and thus "proactive."

162 "coalitionary" proactive aggression: Justin M. Carré, Cheryl M. McCormick, and Ahmad R. Hariri, "The Social Neuroendocrinology of Human Aggression," *Psychoneuroendocrinology* 36, no. 7 (2011): 935–44; and Wrangham, *The Goodness Paradox.*

162 nervous system in proactive aggression: See Wenfeng Zhu, Xiaolin Zhou, and Ling-Xiang Xia, "Brain Structures and Functional Connectivity Associated with Individual Differences in Trait Proactive Aggression," *Scientific Reports* 9, no. 1 (2019): 1–12; Jilly Naaijen, Leandra M. Mulder, Shahrzad Ilbegi et al., "Specific Cortical and Subcortical Alterations for Reactive and Proactive Aggression in Children and Adolescents with Disruptive Behavior," *Neuroimage: Clinical* 27 (2020): 102344; and Meghan E. Flanigan and Scott J. Russo, "Recent Advances in the Study of Aggression," *Neuropsychopharmacology* 44, no. 2 (2019): 241–44.

162 how to extrapolate from behavior in these artificial environments: Mark A. Schmuckler, "What Is Ecological Validity? A Dimensional Analysis," *Infancy* 2, no. 4 (2001): 419–36.

163 Men commit 90 to 95 percent of all murders: United Nations Office on Drugs and Crime, "Global Study on Homicide 2019," Booklet 1: Executive Summary, 2019, 22, https://www.unodc.org/unodc/en/data-and-analysis/global-study-on-homicide.html.

163 When men kill women their motivation is often sexual jealousy: Evolutionary explanations for men's intimate partner violence: James Alan Fox and Emma E. Fridel, "Gender Differences in Patterns and Trends

in US Homicide, 1976–2015," *Violence and Gender* 4, no. 2 (2017): 37–43; and Margo Wilson and Martin Daly, "Coercive Violence by Human Males Against Their Female Partners," in *Sexual Coercion in Primates and Humans: An Evolutionary Perspective on Male Aggression Against Females*, ed. Martin N. Muller and Richard W. Wrangham, 271–91 (Cambridge, MA: Harvard University Press, 2009).

Women almost never kill other women. Like men, women's opposite-sex victims are often their intimate partners. Female homicide rates vary little from country to country compared to the male rates. There's lots of wiggle room in the downward direction for male crime rates, but women's rates are already so low they can't come down much more. So in countries where murder is very rare, like Singapore and Switzerland, sex differences are smaller than in countries where the murder rate is high, like South Africa and Venezuela. See "Global Study on Homicide 2019," United Nations Office on Drugs and Crime, Booklet 2: Homicide: Extent, Patterns, Trends and Criminal Justice Response, 2019.

164   Rape, in particular: Kirsten J. Russell and Christopher J. Hand, "Rape Myth Acceptance, Victim Blame Attribution and Just World Beliefs: A Rapid Evidence Assessment," *Aggression and Violent Behavior* 37 (2017): 153–60.

165   men commit 80 to 85 percent of violent crimes: Federal Bureau of Investigation, "Table 42: Arrests by Sex," FBI 2018 Crime in the United States, Criminal Justice Information Services Division, n.d., https://ucr.fbi.gov/crime-in-the-u .s/2018/crime-in-the-u.s.-2018/topic-pages/tables/table-42.

165   Worldwide, men commit more fraud than women: Markku Heiskanen and Anni Lietonen, "Crime and Gender: A Study on How Men and Women Are Represented in International Crime Statistics," publication series no. 85, European Institute for Crime Prevention and Control, Helsinki, 2016, 59, https://www .heuni.fi/material/attachments/heuni/reports/Ast1S7Egx/Crime_and_gender_ taitto.pdf. See also United Nations Office on Drugs and Crime, "Global Study on Homicide 2019," Booklet 1: Executive Summary, 2019, 22, https://www .unodc.org/unodc/en/data-and-analysis/global-study-on-homicide.html.

For a comprehensive meta-analysis of sex differences in violence and aggression worldwide, see Archer, "Sex Differences in Aggression in Real-World Settings."

For sex differences in fraud, see Bruce Dorris, *Report to the Nations: 2018 Global Study on Occupational Fraud and Abuse*, Association of Certified Fraud Examiners (2018), https://s3-us-west-2.amazonaws.com/acfepublic /2018-report-to-the-nations.pdf.

165   "There is a difference in the degree to which men and women escalate": John Archer, "The Reality and Evolutionary Significance of Human Psychological Sex Differences," *Biological Reviews* 94, no. 4 (2019): 1389.

165   High rates of male aggression appear: Robert L. Cieri, Steven E. Churchill, Robert G. Franciscus, Jingzhi Tan, and Brian Hare, "Craniofacial Feminization, Social Tolerance, and the Origins of Behavioral Modernity," *Current Anthropology* 55, no. 4 (2014): 419–43.

165   Ancient fossil skulls show evidence: Phillip L. Walker, "A Bioarchaeologi-
        cal Perspective on the History of Violence," *Annual Review of Anthropology*
        30, no. 1 (2001): 587; and Patricia Lambert, "Patterns of Violence in Prehis-
        toric Hunter-Gatherer Societies of Coastal Southern California," in *Troubled
        Times: Violence and Warfare in the Past*, ed. David W. Frayer and Debra L.
        Martin, 87–89 (London: Routledge, 1998).

165   among the hunter-gatherer populations remaining today: Nicole Hess,
        Courtney Helfrecht, Edward Hagen, Aaron Sell, and Barry Hewlett, "Inter-
        personal Aggression Among Aka Hunter-Gatherers of the Central African
        Republic," *Human Nature* 21, no. 3 (2010): 330–54.

166   Men are bigger and stronger than women: Haider J. Warraich and Robert
        M. Califf, "Differences in Health Outcomes Between Men and Women:
        Biological, Behavioral, and Societal Factors," *Clinical Chemistry* 65, no. 1
        (2019): 19–23.

166   Boys, like many other young male primates: For sex differences in play, and
        play and hormones, see Melissa Hines, Mihaela Constantinescu, and Debra
        Spencer, "Early Androgen Exposure and Human Gender Development," *Biol-
        ogy of Sex Differences* 6, no. 3 (2015); Vickie L. Pasterski, Mitchell E. Geffner,
        Caroline Brain, Peter Hindmarsh, Charles Brook, and Melissa Hines, "Pre-
        natal Hormones and Postnatal Socialization by Parents as Determinants of
        Male-Typical Toy Play in Girls with Congenital Adrenal Hyperplasia," *Child
        Development* 76, no. 1 (2005): 264–78; D. Spencer, V. Pasterski, S. Neufeld
        et al., "Prenatal Androgen Exposure and Children's Aggressive Behavior and
        Activity Level," *Hormones and Behavior* 96 (2017): 156–65; Sheri A. Beren-
        baum, "Beyond Pink and Blue: The Complexity of Early Androgen Effects
        on Gender Development," *Child Development Perspectives* 12, no. 1 (2018):
        58–64; and Sheri A. Berenbaum and Adriene M. Beltz, "Sexual Differenti-
        ation of Human Behavior: Effects of Prenatal and Pubertal Organizational
        Hormones," *Frontiers in Neuroendocrinology* 32, no. 2 (2011): 183–200.

166   Men are overrepresented in violent contact sports: Dale C. Spencer, "Narra-
        tives of Despair and Loss: Pain, Injury and Masculinity in the Sport of Mixed
        Martial Arts," *Qualitative Research in Sport, Exercise and Health* 4, no. 1 (2012):
        117–37; and Robert O. Deaner and Brandt A. Smith, "Sex Differences in Sports
        Across 50 Societies," *Cross-Cultural Research* 47, no. 3 (2013): 268–309.

166   Players of violent video games: For a comprehensive review of psychological sex
        differences, see Archer, "The Reality and Evolutionary Significance of Human
        Psychological Sex Differences." On sex differences in video game preferences,
        see Kristen Lucas and John L. Sherry, "Sex Differences in Video Game Play:
        A Communication-Based Explanation," *Communication Research* 31, no. 5
        (2004): 499–523; and Melissa Terlecki, Jennifer Brown, Lindsey Harner-Steciw
        et al., "Sex Differences and Similarities in Video Game Experience, Preferences,
        and Self-Efficacy: Implications for the Gaming Industry," *Current Psychology* 30,
        no. 1 (2011): 22–33. On violent fantasies, see Susan Pollak and Carol Gilligan,
        "Images of Violence in Thematic Apperception Test Stories," *Journal of Person-
        ality and Social Psychology* 42, no. 1 (1982): 159–67; and Limor Goldner, Rachel

Lev-Wiesel, and Guy Simon, "Revenge Fantasies After Experiencing Traumatic Events: Sex Differences," *Frontiers in Psychology* 10 (2019), article 886.

166   more variance among men than women: The size of the difference in the variance of male versus female reproductive success varies greatly by mating system and other factors; the sex difference is highest in cultures that are polygynous (even though a small minority of men are polygynously paired in those societies) or serially monogamous. See Gillian R. Brown, Kevin N. Laland, and Monique Borgerhoff Mulder, "Bateman's Principles and Human Sex Roles," *Trends in Ecology and Evolution* 24, no. 6 (2009): 297–304.

166   in more traditional societies: Brown, Laland, and Borgerhoff Mulder, "Bateman's Principles and Human Sex Roles."

167   And in polygynous societies: Polygynous males, particularly those with a relatively high number of wives, are likely to be reproductive winners. See Mhairi A. Gibson and Ruth Mace, "Polygyny, Reproductive Success and Child Health in Rural Ethiopia: Why Marry a Married Man?," *Journal of Biosocial Science* 39, no. 2 (2007): 287–303. Polygyny is widespread among societies (85 percent of societies are polygynous) but occurs much less frequently: only 7 to 14 percent of men are polygynous. However, polygynous societies have more unmarried, unsettled, and unsatisfied men, who are left without mating prospects because fewer women are available (reproductive losers). This situation is associated with greater rates of male violence of all kinds. Alternatively, monogamous societies are associated with decreased levels of male-male violence, increased gender equality, and increased economic productivity. Reduced T among married men and fathers may be a contributing factor.

167   For a review of how cultural norms around mating and parenting relate to various levels of violence, see Joseph Henrich, Robert Boyd, and Peter J. Richerson, "The Puzzle of Monogamous Marriage," *Philosophical Transactions of the Royal Society B: Biological Sciences* 367, no. 1589 (2012): 657–69.

167   how T modulates aggression in men: Carré, McCormick, and Hariri, "The Social Neuroendocrinology of Human Aggression." Evidence in nonhuman animals suggests that social threats, like those to status, reputation, resources, or mates, activate a pathway that increases the likelihood for reactive aggression in the environment of high T (or its metabolites). Men may very well possess similar neural adaptations that underlie the expression of reactive (or proactive) aggression. See Wrangham, "Two Types of Aggression in Human Evolution."

167   significant factor in explaining sex differences in violent behavior: T levels vary not only as a function of whether a man is partnered or an involved father, but also as a function of cultural norms relating to activity, diet, and paternal investment. More single, higher-T men (more in polygynous societies) is associated with higher levels of violence. For a review of the relevant research and an illustration of how culture and biology are deeply intertwined, see Joseph Henrich, *The Weirdest People in the World: How the West Became Psychologically Peculiar and Particularly Prosperous* (New York: Farrar, Straus and Giroux, 2020), 268–83.

168   T levels of adult male chimps rise to their highest levels when females are in estrus: Martin N. Muller and Richard W. Wrangham, "Dominance, Aggression and Testosterone in Wild Chimpanzees: A Test of the 'Challenge Hypothesis,'" *Animal Behaviour* 67, no. 1 (2004): 113–23.

168   evidence does suggest, however, that men can subconsciously detect subtle changes: Martie G. Haselton and Kelly Gildersleeve, "Can Men Detect Ovulation?," *Current Directions in Psychological Science* 20, no. 2 (2011): 87–92; Geoffrey Miller, Joshua M. Tybur, and Brent D. Jordan, "Ovulatory Cycle Effects on Tip Earnings by Lap Dancers: Economic Evidence for Human Estrus?," *Evolution and Human Behavior* 28, no. 6 (2007): 375–81; Saul L. Miller and Jon K. Maner, "Scent of a Woman: Men's Testosterone Responses to Olfactory Ovulation Cues," *Psychological Science* 21, no. 2 (2010): 276–83; and see Steven W. Gangestad and Martie G. Haselton, "Human Estrus: Implications for Relationship Science," *Current Opinion in Psychology* 1 (2015): 45–51, for a review of relevant shifts in women's physiology and behavior (which may signal ovulation) across the menstrual cycle.

168   why men stick around even when women can't get pregnant: Why human females have evolved "concealed ovulation" where other primates "advertise" theirs is not completely understood. One leading theory is that concealing the time of greatest fertility results in increased survival of our young, which are born relatively helpless, needing high levels of parental care. If a man can't detect (subconsciously) when a woman is ovulating, then that incentivizes him to remain in a relationship with his sexual partner in order to increase his chances of fertilization, while keeping other potential mates away when his mate may be fertile. Men and women both benefit by forming a strong and lasting social and sexual pair bond that increases the male investment in his offspring and, in turn, offspring survival. See David C. Geary and Mark V. Flinn, "Evolution of Human Parental Behavior and the Human Family," *Parenting* 1, no. 1–2 (2001): 5–61. For a review of competing theories of concealed ovulation, see Beverly I. Strassmann, "Sexual Selection, Paternal Care, and Concealed Ovulation in Humans," *Ethology and Sociobiology* 2, no. 1 (1981): 31–40.

169   many men invest time and energy in their children: Ryan Schacht, Helen E. Davis, and Karen L. Kramer, "Patterning of Paternal Investment in Response to Socioecological Change," *Frontiers in Ecology and Evolution* 6 (2018), article 142.

169   a human dad's T levels tend to drop: Whether T falls in men who are romantically paired or new fathers depends on many factors, including investment in the mating and paternal relationships, and is a small to medium effect: see Nicholas M. Grebe, Ruth E. Sarafin, Chance R. Strenth, and Samuele Zilioli, "Pair-Bonding, Fatherhood, and the Role of Testosterone: A Meta-Analytic Review," *Neuroscience and Biobehavioral Reviews* 98 (2019): 221–33. Also reviewed in Peter B. Gray, Timothy S. McHale, and Justin M. Carré, "A Review of Human Male Field Studies of Hormones and Behavioral Reproductive Effort," *Hormones and Behavior* 91 (2017): 52–67. For more information,

see Lee T. Gettler, Thomas W. McDade, Alan B. Feranil, and Christopher W. Kuzawa, "Longitudinal Evidence That Fatherhood Decreases Testosterone in Human Males," *Proceedings of the National Academy of Sciences* 108, no. 39 (2011): 16194–99; and Christopher W. Kuzawa, Lee T. Gettler, Martin N. Muller, Thomas W. McDade, and Alan B. Feranil, "Fatherhood, Pairbonding and Testosterone in the Philippines," *Hormones and Behavior* 56, no. 4 (2009): 429–35. T levels are higher in romantically paired men who have a greater interest in cheating on their partner. See Matthew McIntyre, Steven W. Gangestad, Peter B. Gray et al., "Romantic Involvement Often Reduces Men's Testosterone Levels—But Not Always: The Moderating Role of Extrapair Sexual Interest," *Journal of Personality and Social Psychology* 91, no. 4 (2006): 642–51.

169   decline in T helps direct attention away from rivals: For a comprehensive review of how testosterone mediates "life history" trade-offs, that is, growth, maintenance, and reproduction (involving mating and parenting), see Richard G. Bribiescas, "Reproductive Ecology and Life History of the Human Male," *American Journal of Physical Anthropology* 116, no. S33 (2001): 148–76.

170   After all, those at the top are winning: Benefits of high social status and motivations for achieving: Joey T. Cheng, Jessica L. Tracy, and Joseph Henrich, "Pride, Personality, and the Evolutionary Foundations of Human Social Status," *Evolution and Human Behavior* 31, no. 5 (2010): 334–47; and Christopher Von Rueden, Michael Gurven, and Hillard Kaplan, "Why Do Men Seek Status? Fitness Payoffs to Dominance and Prestige," *Proceedings of the Royal Society B: Biological Sciences* 278, no. 1715 (2011): 2223–32. Human males are more efficient at resolving conflicts than females, possibly facilitated by larger male groups with stricter dominance hierarchies: Joyce F. Benenson and Richard W. Wrangham, "Cross-Cultural Sex Differences in Post-Conflict Affiliation Following Sports Matches," *Current Biology* 26, no. 16 (2016): 2208–12; and Chris Von Rueden, Sarah Alami, Hillard Kaplan, and Michael Gurven, "Sex Differences in Political Leadership in an Egalitarian Society," *Evolution and Human Behavior* 39, no. 4 (2018): 402–11.

170   On average, these camps consisted of: Hunter-gatherer population sizes: Wrangham, *The Goodness Paradox*, 154–55; and Frank W. Marlowe, "Hunter-Gatherers and Human Evolution," *Evolutionary Anthropology* 14, no. 2 (2005): 54–67.

170   everyone in a given society knew everyone else: Knowledge of others in the wider society: Kim R. Hill, Brian M. Wood, Jacopo Baggio, A. Magdalena Hurtado, and Robert T. Boyd, "Hunter-Gatherer Inter-Band Interaction Rates: Implications for Cumulative Culture," *PloS One* 9, no. 7 (2014): e102806.

171   T levels in men normally change throughout the day: Percent in T decline across the day: Michael J. Diver, Komal E. Imtiaz, Aftab M. Ahmad, Jiten P. Vora, and William D. Fraser, "Diurnal Rhythms of Serum Total, Free and Bioavailable Testosterone and of SHBG in Middle-Aged Men Compared with Those in Young Men," *Clinical Endocrinology* 58, no. 6 (2003): 710–17.

172   A soccer or boxing match might not seem at first glance: Robert O. Deaner, Shea M. Balish, and Michael P. Lombardo, "Sex Differences in Sports Interest and Motivation: An Evolutionary Perspective," *Evolutionary Behavioral Sciences* 10, no. 2 (2016): 73.

172   It was all part of a simple experiment: Paul C. Bernhardt, James M. Dabbs Jr., Julie A. Fielden, and Candice D. Lutter, "Testosterone Changes During Vicarious Experiences of Winning and Losing Among Fans at Sporting Events," *Physiology and Behavior* 65, no. 1 (1998): 59–62.

173   Many factors affect how T responds to competition: John C. Wingfield, Marilyn Ramenofsky, Robert E. Hegner, and Gregory F. Ball, "Whither the Challenge Hypothesis?," *Hormones and Behavior* 123 (2020): 104588; and Donna L. Maney, "The Challenge Hypothesis: Triumphs and Caveats," *Hormones and Behavior* 123 (2020): 104663. For a review of the relevant literature, see Joe Herbert, *Testosterone: Sex, Power, and the Will to Win* (New York: Oxford University Press, 2015), 109–29.

173   in nonhuman species, the effect clearly is mediated by testosterone: Rui F. Oliveira, Marco Lopes, Luis A. Carneiro, and Adelino V. M. Canário, "Watching Fights Raises Fish Hormone Levels," *Nature* 409, no. 6819 (2001): 475.

173   This effect lasts about a month in Syrian hamsters: M. B. Solomon, M. C. Karom, A. Norvelle, C. A. Markham, W. D. Erwin, and K. L. Huhman, "Gonadal Hormones Modulate the Display of Conditioned Defeat in Male Syrian Hamsters," *Hormones and Behavior* 56, no. 4 (2009): 423–28.

174   the effects of experimentally elevated T levels: Oliver C. Schultheiss, Kenneth L. Campbell, and David C. McClelland, "Implicit Power Motivation Moderates Men's Testosterone Responses to Imagined and Real Dominance Success," *Hormones and Behavior* 36, no. 3 (1999): 234–41; and Shawn N. Geniole and Justin M. Carré, "Human Social Neuroendocrinology: Review of the Rapid Effects of Testosterone," *Hormones and Behavior* 104 (2018): 192–205.

176   The span of DNA that comprises each androgen receptor gene has something in it called a "CAG repeat": Christoph Eisenegger, Robert Kumsta, Michael Naef, Jörg Gromoll, and Markus Heinrichs, "Testosterone and Androgen Receptor Gene Polymorphism Are Associated with Confidence and Competitiveness in Men," *Hormones and Behavior* 92 (2017): 93–102.

176   one's CAG repeat turns out to be associated with all sorts of things: Merlin G. Butler and Ann M. Manzardo, "Androgen Receptor (AR) Gene CAG Trinucleotide Repeat Length Associated with Body Composition Measures in Non-Syndromic Obese, Non-Obese and Prader-Willi Syndrome Individuals," *Journal of Assisted Reproduction and Genetics* 32, no. 6 (2015): 909–15.

177   they will choose the side in which they previously received a nice dose of testosterone: M. G. Packard, A. H. Cornell, and G. M. Alexander, "Rewarding Affective Properties of Intra-Nucleus Accumbens Injections of Testosterone," *Behavioral Neuroscience* 111, no. 1 (1997): 219–24; and Jeffrey Parrilla-Carrero, Orialis Figueroa, Alejandro Lugo et al., "The Anabolic

Steroids Testosterone Propionate and Nandrolone, but Not 17alpha-Methyltestosterone, Induce Conditioned Place Preference in Adult Mice," *Drug and Alcohol Dependence* 100, no. 1–2 (2009): 122–27.

177 The parts of the brain that are rich in the neurotransmitter dopamine: Tertia D. Purves-Tyson, Samantha J. Owens, Kay L. Double, Reena Desai, David J. Handelsman, and Cynthia S. Weickert, "Testosterone Induces Molecular Changes in Dopamine Signaling Pathway Molecules in the Adolescent Male Rat Nigrostriatal Pathway," *PloS One* 9, no. 3 (2014): e91151; and Cheryl A. Frye, "Some Rewarding Effects of Androgens May Be Mediated by Actions of Its 5α-Reduced Metabolite 3α-Androstanediol," *Pharmacology, Biochemistry, and Behavior* 86, no. 2 (2007): 354–67.

177 T actually increases the amount of dopamine: M. A. de Souza Silva, C. Mattern, B. Topic, T. E. Buddenberg, and J. P. Huston, "Dopaminergic and Serotonergic Activity in Neostriatum and Nucleus Accumbens Enhanced by Intranasal Administration of Testosterone," *European Neuropsychopharmacology* 19, no. 1 (2009): 53–63.

177 increasing levels of T appear to be motivating and rewarding: Shawn N. Geniole, Tanya L. Procyshyn, Nicole Marley et al., "Using a Psychopharmacogenetic Approach to Identify the Pathways Through Which—and the People for Whom—Testosterone Promotes Aggression," *Psychological Science* 30, no. 4 (2019): 481–94.

177 it's complicated: Robert M. Sapolsky, *The Trouble with Testosterone: And Other Essays on the Biology of the Human Predicament* (New York: Scribner, 1998).

177 T has other effects that make severe aggression more likely: Baris O. Yildirim and Jan J. L. Derksen, "A Review on the Relationship Between Testosterone and the Interpersonal/Affective Facet of Psychopathy," *Psychiatry Research* 197, no. 3 (2012): 181–98.

177 Reducing T reverses these effects: Justin M. Carré, Susan K. Putnam, and Cheryl M. McCormick, "Testosterone Responses to Competition Predict Future Aggressive Behaviour at a Cost to Reward in Men," *Psychoneuroendocrinology* 34, no. 4 (2009): 561–70.

177 fear and the perception of pain are decreased: Katy Vincent, Catherine Warnaby, Charlotte J. Stagg, Jane Moore, Stephen Kennedy, and Irene Tracy, "Brain Imaging Reveals That Engagement of Descending Inhibitory Pain Pathways in Healthy Women in a Low Endogenous Estradiol State Varies with Testosterone," *Pain* 154, no. 4 (2013): 515–24; and J. C. Choi, Y.-H. Park, S. K. Park et al., "Testosterone Effects on Pain and Brain Activation Patterns," *Acta Anaesthesiologica Scandinavica* 61, no. 6 (2017): 668–75.

177 experiment in a captive group of Talapoin monkeys: A. F. Dixson and J. Herbert, "Testosterone, Aggressive Behavior and Dominance Rank in Captive Adult Male Talapoin Monkeys (*Miopithecus talapoin*)," *Physiology and Behavior* 18, no. 3 (1977): 539–43.

178 "enough to grow antlers and a beard on every neuron in his brain": Sapolsky, *The Trouble with Testosterone*, 154.

178		if you shot a bunch of Buddhist monks up with testosterone: Kim Post, "Sapolsky Gives Lecture on Violence, Human Behavior," *Triangle*, Drexel University student newspaper, April 21, 2017, https://www.thetriangle.org /news/sapolsky-gives-lecture-violence-human-behavior/.

178		That LH, in turn, needs about one hour: N. A. Bridges, P. C. Hindmarsh, P. J. Pringle, D. R. Matthews, and C. G. D. Brook, "The Relationship Between Endogenous Testosterone and Gonadotrophin Secretion," *Clinical Endocrinology* 38, no. 4 (1993): 373–78.

178		Research suggests one possibility: Robert M. Sapolsky, "Stress-Induced Elevation of Testosterone Concentrations in High Ranking Baboons: Role of Catecholamines," *Endocrinology* 118, no. 4 (1986): 1630–35; and Kathleen V. Casto and David A. Edwards, "Testosterone, Cortisol, and Human Competition," *Hormones and Behavior* 82 (2016): 21–37.

179		T could have important, and more rapid, "nongenomic" effects: C. D. Foradori, M. J. Weiser, and R. J. Handa, "Non-Genomic Actions of Androgens," *Frontiers in Neuroendocrinology* 29, no. 2 (2008): 169–81; and Cynthia A. Heinlein and Chawnshang Chang, "The Roles of Androgen Receptors and Androgen-Binding Proteins in Nongenomic Androgen Actions," *Molecular Endocrinology* 16, no. 10 (2002): 2181–87.

179		T levels in some nonhuman female animals: See chapter 6 for discussion of T in aggression in female nonhuman animals.

180		this might simply be a by-product of the adrenals responding to stress: Some researchers are concerned that women are largely left out of these studies, and they are right that there are far more that examine this effect in men. Some of this apparent neglect is likely because previous attempts to find the "winner-loser" effect in women have mostly failed. Research on women's T is complicated, partly because of previously discussed difficulties in measuring women's T, and women's T levels change as a function of menstrual cycles and birth control status, so these must be taken into account; and researchers want to obtain positive results so they can get their studies published. (I'm not endorsing leaving women out, but there are understandable reasons for their reduced inclusion.)

180		Of the studies on the winner-loser effect that include women: Shawn N. Geniole, Brian M. Bird, Erika L. Ruddick, and Justin M. Carré, "Effects of Competition Outcome on Testosterone Concentrations in Humans: An Updated Meta-Analysis," *Hormones and Behavior* 92 (2017): 37–50; and K. V. Casto, D. A. Edwards, M. Akinola, C. Davis, and P. H. Mehta, "Testosterone Reactivity to Competition and Competitive Endurance in Men and Women," *Hormones and Behavior* 123 (2020): 104655.

180		it would be astounding if men's and women's testosterone production responded in the same way to competition: Casto et al., "Testosterone Reactivity to Competition."

180		There are other hormones involved in competition: E. Barel, S. Shahrabani, and O. Tzischinsky, "Sex Hormone/Cortisol Ratios Differentially Modulate Risk-Taking in Men and Women," *Evolutionary Psychology* 15, no. 1 (2017):

1–10; and Pranjal H. Mehta, Amanda C. Jones, and Robert A. Josephs, "The Social Endocrinology of Dominance: Basal Testosterone Predicts Cortisol Changes and Behavior Following Victory and Defeat," *Journal of Personality and Social Psychology* 94, no. 6 (2008): 1078–93.

181   "The police gave me a subtle nod of approval": Fairless, *Mad Blood Stirring*, 1.

181   "From an early age, small boys were taught to think much of their own honor": Quoted in R. E. Nisbett, *Culture of Honor: The Psychology of Violence in the South* (Boulder, CO: Westview, 1996; Abingdon, UK: Taylor and Francis, 2018), 2.

181   As Steven Pinker has documented in *The Better Angels of Our Nature*: Steven Pinker, *The Better Angels of Our Nature: Why Violence Has Declined* (New York: Penguin, 2012), ch. 3, 104.

CHAPTER 8

183   "Tell Mrs. Coolidge": James R. Wilson, Robert E. Kuehn, and Frank A. Beach, "Modification in the Sexual Behavior of Male Rats Produced by Changing the Stimulus Female," *Journal of Comparative and Physiological Psychology* 56, no. 3 (1963): 636.

183   Researchers have demonstrated it in a number of animals: More on the Coolidge effect: David M. Buss, *The Evolution of Desire*, rev. ed. (New York: Basic Books, 2003), 80; also Susan M. Hughes, Toe Aung, Marissa A. Harrison, Jack N. LaFayette, and Gordon G. Gallup Jr., "Experimental Evidence for Sex Differences in Sexual Variety Preferences: Support for the Coolidge Effect in Humans," *Archives of Sexual Behavior* (May 21, 2020), https://doi.org/10.1007/s10508-020-01730-x.

184   Dopamine has many functions: James G. Pfaus, "Dopamine: Helping Males Copulate for at Least 200 Million Years: Theoretical Comment on Kleitz-Nelson et al. (2010)," *Behavioral Neuroscience* 124, no. 6 (2010): 877–80.

184   Bang, up jump his dopamine levels: M. Dean Graham and James G. Pfaus, "Differential Regulation of Female Sexual Behaviour by Dopamine Agonists in the Medial Preoptic Area," *Pharmacology, Biochemistry, and Behavior* 97, no. 2 (2010): 284–92.

184   The neural systems (including dopamine changes) that motivate the pursuit of sex and the reward: Catriona Wilson, George C. Nomikos, Maria Collu, and Hans C. Fibiger, "Dopaminergic Correlates of Motivated Behavior: Importance of Drive," *Journal of Neuroscience* 15, no. 7 (1995): 5169–78.

185   When she can control the pace of mating: Raúl G. Paredes and Berenice Vazquez, "What Do Female Rats Like About Sex? Paced Mating," *Behavioural Brain Research* 105, no. 1 (1999): 117–27.

185   the motivation he needs to pursue her, mate, and ejaculate again: Dennis F. Fiorino, Ariane Coury, and Anthony G. Phillips, "Dynamic Changes in Nucleus Accumbens Dopamine Efflux During the Coolidge Effect in Male Rats," *Journal of Neuroscience* 17, no. 12 (1997): 4849–55.

185   His penis must become erect: For a review of the roles of hormones and neurotransmitters in coordinating motivation and movements necessary for mating, focusing on coordinating the slow effects of T with the faster effects of dopamine and serotonin: Elaine M. Hull, John W. Muschamp, and Satoru Sato, "Dopamine and Serotonin: Influences on Male Sexual Behavior," *Physiology and Behavior* 83, no. 2 (2004): 291–307.

185   Dopamine is crucial for the expression of all kinds of motivated behavior: Pfaus, "Dopamine: Helping Males Copulate."

186   But if another fertile female comes around: High T preps the nervous system in prenatal development and in puberty, so that dopamine rises in the right parts of the brain at the right time, motivating male animals to pursue sexual stimuli in adulthood. T and dopamine together also help to ensure that all the steps on the road to sex, not just the destination, are rewarding. Testosterone affects neural circuitry so that dopamine levels are increased in an area of the brain central to sexual behavior (the medial preoptic area, or MPOA) in the presence of relevant sexual stimuli. When a male encounters the sights or smell of a fertile female, his high T leads to increased dopamine in the MPOA, making it more likely that he will make the movements necessary to investigate, pursue, and mate with the female. And he'll like doing it! These effects of T on the nervous system don't vanish the moment that T leaves the blood. In fact, they can last for several weeks or more, depending on the species. If a male rat is castrated, within a few weeks he will lose interest in mating. But if you squirt some dopamine into the right brain areas while he's in the presence of an estrous female, even without recent testosterone exposure, he'll once again respond to her signals and show every interest in mating. Testosterone appears to set the neural stage for dopamine to do its thing, and motivate and reward sexual behavior in general, the seeking out of new sexual partners in particular. See Margaret R. Bell and Cheryl L. Sisk, "Dopamine Mediates Testosterone-Induced Social Reward in Male Syrian Hamsters," *Endocrinology* 154, no. 3 (2013): 1225–34.

186   when it comes to having sex, men want to do it more often: John Archer, "The Reality and Evolutionary Significance of Human Psychological Sex Differences," *Biological Reviews* 94, no. 4 (2019): 1381–415.

187   this conversion happens in "peripheral tissues": L. Liu, J. Kang, X. Ding, D. Chen, Y. Zhou, and H. Ma, "Dehydroepiandrosterone-Regulated Testosterone Biosynthesis via Activation of the Erk1/2 Signaling Pathway in Primary Rat Leydig Cells," *Cellular Physiology and Biochemistry* 36, no. 5 (2015): 1778–92.

187   It is this adrenally derived T: Athanasios Antoniou-Tsigkos, Evangelia Zapanti, Lucia Ghizzoni, and George Mastorakos, "Adrenal Androgens," *EndoText*, January 5, 2019, https://www.ncbi.nlm.nih.gov/books/NBK278929/. Adrenal androgens in men contribute about 5 percent of total T and thus aren't an important contributor to masculinizing effects. But in women they provide a substantial portion of total T, as much as two-thirds, depending on cycle phase. The portion of T contributed by adrenal androgens is reduced during

midcycle (to about 40 percent), when the amount of T released into circulation by the ovaries increases.

187 adrenal androgens aren't enough to stimulate the growth spurt: Benjamin C. Campbell, "Adrenarche and Middle Childhood," *Human Nature* 22, no. 3 (2011): 327.

187 These changes in bodies and feelings aren't just happening in Western cultures: Peter B. Gray, "Evolution and Human Sexuality," *American Journal of Physical Anthropology* 152 (2013): 94–118.

187 Girls enter puberty about a year before boys: These ages are from U.S. data; sex differences in age hold across cultures, but exact ages vary. Data on pubertal timing from non-Western cultures show high levels of natural variation. For example, see Rebecca Sear, Paula Sheppard, and David A. Coall, "Cross-Cultural Evidence Does Not Support Universal Acceleration of Puberty in Father-Absent Households," *Philosophical Transactions of the Royal Society B* 374, no. 1770 (2019): 20180124.

187 A longer period of juvenile growth gives young males time: On earlier puberty in girls, and the evolutionary explanations: Natalie V. Motta-Mena and David A. Puts, "Endocrinology of Human Female Sexuality, Mating, and Reproductive Behavior," *Hormones and Behavior* 91 (2017): 19–35.

188 adolescent boys and girls report having a crush on someone: J. Dennis Fortenberry, "Puberty and Adolescent Sexuality," *Hormones and Behavior* 64, no. 2 (2013): 280–87; and Margaret R. Bell, "Comparing Postnatal Development of Gonadal Hormones and Associated Social Behaviors in Rats, Mice, and Humans," *Endocrinology* 159, no. 7 (2018): 2596–613.

188 by the age of eighteen the majority of teens will have had sex: Adolescent trends in sexual behavior vary significantly by socioeconomic status, ethnicity, and cultural norms. Stephen T. Russell, "Conceptualizing Positive Adolescent Sexuality Development," *Sexuality Research and Social Policy* 2, no. 3 (2005): 4.

188 coordinate large and often relatively slow changes: Peter T. Ellison, "Endocrinology, Energetics, and Human Life History: A Synthetic Model," *Hormones and Behavior* 91 (2017): 97–106.

188 In girls, T levels barely rise: Changes in T and E levels from prenatal through the end of puberty in humans and nonhuman animals: Bell, "Comparing Postnatal Development of Gonadal Hormones." Hormone levels on p. 2598.

189 during puberty that these previously masculinized brain structures are again exposed to high levels of T: Testosterone in male puberty is thought to be a second kind of organizational period, in which neural structures that were "organized" perinatally are further sculpted. These structures are the ones T acts on specifically to activate male sexual behavior. See Kalynn M. Schulz, Heather A. Molenda-Figueira, and Cheryl L. Sisk, "Back to the Future: The Organizational–Activational Hypothesis Adapted to Puberty and Adolescence," *Hormones and Behavior* 55, no. 5 (2009): 597–604.

190 homosexual behavior was thought to be masculine: Ruth Mazo Karras, "Active/Passive, Acts/Passions: Greek and Roman Sexualities," *American Historical Review* 105, no. 4 (2000): 1250–65.

190　homosexuality is illegal and punishable by death: Max Bearak and Darla Cameron, "Here Are the 10 Countries Where Homosexuality May Be Punished by Death," *Washington Post*, June 16, 2016.

190　young women are likely to be "slut-shamed": Joyce J. Endendijk, Anneloes L. van Baar, and Maja Deković, "He Is a Stud, She Is a Slut! A Meta-Analysis on the Continued Existence of Sexual Double Standards," *Personality and Social Psychology Review* 24, no. 2 (2020): 163–90; Derek A. Kreager and Jeremy Staff, "The Sexual Double Standard and Adolescent Peer Acceptance," *Social Psychology Quarterly* 72, no. 2 (2009): 143–64.

190　most successful men in most hunter-gatherer societies sire: The most successful men among farmers are far more successful than the least successful, but this variation is much smaller than that among hunter-gatherers. See Laura Betzig, "Means, Variances, and Ranges in Reproductive Success: Comparative Evidence," *Evolution and Human Behavior* 33, no. 4 (2012): 309–17.

190　The prize for superlative achievement in the fatherhood: Ewen Callaway, "Genghis Khan's Genetic Legacy Has Competition," *Nature*, January 23, 2015.

190　Genghis's sons inherited his enthusiasm for promiscuity: Razib Khan, "1 in 200 Men Are Direct Descendants of Genghis Khan," *Discover*, August 5, 2010. See also Shao-Qing Wen et al., "Molecular Genealogy of Tusi Lu's Family Reveals Their Paternal Relationship with Jochi, Genghis Khan's Eldest Son," *Journal of Human Genetics* 64, no. 8 (2019): 815–20.

191　"Big Dad" recently died: Ny MaGee, "Popular Angolan Polygamist Who Had 156 Children from 49 Wives Dies at 73," Lee Bailey's Eurweb, May 1, 2020, https://eurweb.com/2020/05/01/popular-angolan-polygamist-who-had-156-children-from-49-wives-dies-at-73/.

191　a human baby's chances are much better if he stays to help out: For a review of the circumstances under which male parental care may evolve, see David C. Geary, *Male, Female: The Evolution of Human Sex Differences*, 3rd ed. (Washington, DC: American Psychological Association, 2021), 83–88.

191　a better reproductive bet than increasing her number of sexual partners: For reviews of the complexities and varieties of women's mating strategies, see Elizabeth Cashdan, "Women's Mating Strategies," *Evolutionary Anthropology: Issues, News, and Reviews* 5, no. 4 (1996): 134–43; and Steven W. Gangestad and Jeffry A. Simpson, "Toward an Evolutionary History of Female Sociosexual Variation," *Journal of Personality* 58, no. 1 (1990): 69–96.

192　women tend to prefer mates who don't just have high social status: David M. Buss and David P. Schmitt, "Mate Preferences and Their Behavioral Manifestations," *Annual Review of Psychology* 70 (2019): 77–110; Archer, "The Reality and Evolutionary Significance of Human Psychological Sex Differences"; and J. Michael Bailey, Steven Gaulin, Yvonne Agyei, and Brian A. Gladue, "Effects of Gender and Sexual Orientation on Evolutionarily Relevant Aspects of Human Mating Psychology," *Journal of Personality and Social Psychology* 66, no. 6 (1994): 1081.

192　polygyny (many wives) is much more common than polyandry (many husbands): Ryan Schacht and Karen L. Kramer, "Are We Monogamous? A Review of the Evolution of Pair-Bonding in Humans and Its Contemporary Variation Cross-Culturally," *Frontiers in Ecology and Evolution* 7, no. 230 (2019).

193   enthusiasm for sex and taste for variety: Steve Stewart-Williams, *The Ape That Understood the Universe: How the Mind and Culture Evolve* (Cambridge: Cambridge University Press, 2018), 75–77. The size of sex differences in human sexuality is relatively small, at least when compared to those in most other species, the evolutionary explanation being that human babies are costly to produce, being relatively large investments of time and energy, and survival is increased when men help out. Paternal investment results in reduced sex differences in traits like aggression and mating competition. Men and women have many overlapping desires when it comes to sex and relationships: they both seek long-term relationships with attractive partners, sometimes cheat on those mates, masturbate, and view pornography.

The existence of a sex difference in the preference for casual sex does not mean that women don't also express the desire (and the behavior) for it. Casual sex can be part of an adaptive suite of mating strategies for women. Sarah Hrdy, a pioneering evolutionary biologist and primatologist, has devoted her career to understanding the role that females have played in driving evolutionary change over time. Hrdy explains that a woman's preference for casual versus committed sex "depends not just on her sex, or her 'essential' nature, but on her ecological, demographic, historical, and even more her immediate endocrinological circumstances and the various options open to her." Sarah Blaffer Hrdy, *The Woman That Never Evolved* (Cambridge, MA: Harvard University Press, 1999), xxiii. Others have also observed that having multiple sexual partners may bring reproductive benefits, for example, by making it easier to "trade up" for someone who is a better provider, or gaining more resources from sex partners that ultimately benefit her offspring. See, for example, Bailey et al., "Effects of Gender and Sexual Orientation on Evolutionarily Relevant Aspects of Human Mating Psychology"; and Heidi Greiling and David M. Buss, "Women's Sexual Strategies: The Hidden Dimension of Extra-Pair Mating," *Personality and Individual Differences* 28, no. 5 (2000): 929–63.

193   in a 2009 study, psychologist Richard Lippa and his team analyzed data: Richard A. Lippa, "Sex Differences in Sex Drive, Sociosexuality, and Height Across 53 Nations: Testing Evolutionary and Social Structural Theories," *Archives of Sexual Behavior* 38, no. 5 (2009): 631–51. Lippa's findings replicated those of another large cross-cultural study: David P. Schmitt, "Universal Sex Differences in the Desire for Sexual Variety: Tests from 52 Nations, 6 Continents, and 13 Islands," *Journal of Personality and Social Psychology* 85, no. 1 (2003): 85.

194   To give you a better idea of the size of the difference: The size of group differences can be measured in a number of ways, including the popular "Cohen's d," and also the more intuitive "common language effect size" ("CL"), which I've used here. CL measures group differences in terms of the probability that any individual in a population will possess a given trait. See Lippa, "Sex

Differences in Sex Drive"; and Stewart-Williams, *The Ape That Understood the Universe*, 75–79.

195   These results are the tip of an iceberg of evidence: For a review of sex differences in preferences for casual sex and evidence for robustness of cross-cultural findings, see Geary, *Male, Female*, 203–7. For similar findings to Lippa, "Sex Differences in Sex Drive," on robust cross-cultural sex differences in sociosexuality, see Schmitt, "Universal Sex Differences in the Desire for Sexual Variety"; and Lee Ellis, "Identifying and Explaining Apparent Universal Sex Differences in Cognition and Behavior," *Personality and Individual Differences* 51, no. 5 (2011): 552–61. Also see Bailey et al., "Effects of Gender and Sexual Orientation on Evolutionarily Relevant Aspects of Human Mating Psychology."

195   more creative methods: Marco Del Giudice, David A. Puts, David C. Geary, and David P. Schmitt, "Sex Differences in Brain and Behavior: Eight Counterpoints," *Psychology Today*, April 8, 2019, https://www.psychologytoday.com/us/blog/sexual-personalities/201904/sex-differences-in-brain-and-behavior-eight-counterpoints; and David P. Schmitt, "Can We Trust What Men and Women Reveal in Sex Surveys?," *Psychology Today*, July 11, 2017, https://www.psychologytoday.com/us/blog/sexual-personalities/201707/can-we-trust-what-men-and-women-reveal-sex-surveys.

195   what the penis says: The penile plethysmograph results correlate highly with subjective sexual arousal; the vaginal photoplethysmograph reflects female sexual arousal, although it is less correlated with reports of subjective arousal. Kelly D. Suschinsky, Martin L. Lalumière, and Meredith L. Chivers, "Sex Differences in Patterns of Genital Sexual Arousal: Measurement Artifacts or True Phenomena?," *Archives of Sexual Behavior* 38, no. 4 (2009): 559–73.

196   a hearty sexual response returns in men: For a review of evidence for the Coolidge effect in humans (as close as you can get), see Hughes et al., "Experimental Evidence for Sex Differences in Sexual Variety Preferences."

196   If the effect is present in women, it is much weaker: Hughes et al., "Experimental Evidence for Sex Differences in Sexual Variety Preferences"; Elisa Ventura-Aquino, Alonso Fernández-Guasti, and Raúl G. Paredes, "Hormones and the Coolidge Effect," *Molecular and Cellular Endocrinology* 467 (2018): 42–48.

196   Men are the main consumers of pornography: For an overview of evolutionary origins of the male preference for visual sexual stimuli and a discussion of its contents from this perspective, see Donald Symons, *The Evolution of Human Sexuality* (New York: Oxford University Press, 1979), 170–84.

196   Men are almost exclusively clients of prostitutes: For men's greater preference for sex without strings, see Richard A. Lippa, "The Preferred Traits of Mates in a Cross-National Study of Heterosexual and Homosexual Men and Women: An Examination of Biological and Cultural Influences," *Archives of Sexual Behavior* 36, no. 2 (2007): 193–208; J. Michael Bailey, *The Man Who Would Be Queen: The Science of Gender-Bending and Transsexualism* (Washington, DC: Joseph Henry Press, 2003), 92; Stewart-Williams, *The Ape That Understood the Universe*, 78–84. On sex differences in use of infidelity

websites: Jana Hackathorn and Brien K. Ashdown, "The Webs We Weave: Predicting Infidelity Motivations and Extradyadic Relationship Satisfaction," *Journal of Sex Research* (April 6, 2020): 1–13.

197   those measures all decline for men who shut down their T production: Effects of T suppression for medical treatment: Evan Ng, Henry H. Woo, Sandra Turner et al., "The Influence of Testosterone Suppression and Recovery on Sexual Function in Men with Prostate Cancer: Observations from a Prospective Study in Men Undergoing Intermittent Androgen Suppression," *Journal of Urology* 187, no. 6 (2012): 2162–67. The effects of changing T levels in transgender people are covered in chapter 9; for a review, see Mats Holmberg, Stefan Arver, and Cecilia Dhejne, "Supporting Sexuality and Improving Sexual Function in Transgender Persons," *Nature Reviews Urology* 16, no. 2 (2019): 121–39.

197   A man's T responds to physiological, social, and environmental circumstances: Peter B. Gray, Timothy S. McHale, and Justin M. Carré, "A Review of Human Male Field Studies of Hormones and Behavioral Reproductive Effort," *Hormones and Behavior* 91 (2017): 52–67.

198   When the baby comes, T may drop even more: Anne E. Storey, Carolyn J. Walsh, Roma L. Quinton, and Katherine E. Wynne-Edwards, "Hormonal Correlates of Paternal Responsiveness in New and Expectant Fathers," *Evolution and Human Behavior* 21, no. 2 (2000): 79–95; Peter B. Gray, J. C. Parkin, and M. E. Samms-Vaughan, "Hormonal Correlates of Human Paternal Interactions: A Hospital-Based Investigation in Urban Jamaica," *Hormones and Behavior* 52, no. 4 (2007): 499–507; Lee T. Gettler, Patty X. Kuo, and Sonny Agustin Bechayda, "Fatherhood and Psychobiology in the Philippines: Perspectives on Joint Profiles and Longitudinal Changes of Fathers' Estradiol and Testosterone," *American Journal of Human Biology* 30, no. 6 (2018): e23150.

198   how much T changes in a new dad depends heavily on his culture: For a comprehensive review of neuroendocrine correlates of responses to parenting in men, including the potential involvement of oxytocin, vasopressin, cortisol, and testosterone, see Sari M. van Anders, Richard M. Tolman, and Gayatri Jainagaraj, "Examining How Infant Interactions Influence Men's Hormones, Affect, and Aggression Using the Michigan Infant Nurturance Simulation Paradigm," *Fathering* 12, no. 2 (2014): 143.

198   T levels of the Datoga dads were no different: Martin N. Muller, Frank W. Marlowe, Revocatus Bugumba, and Peter T. Ellison, "Testosterone and Paternal Care in East African Foragers and Pastoralists," *Proceedings of the Royal Society B: Biological Sciences* 276, no. 1655 (2009): 347–54.

198   a dad's focus on the family: Peter B. Gray, Chi-Fu Jeffrey Yang, and Harrison G. Pope Jr., "Fathers Have Lower Salivary Testosterone Levels Than Unmarried Men and Married Non-Fathers in Beijing, China," *Proceedings of the Royal Society B: Biological Sciences* 273, no. 1584 (2006): 333–39; T drops in new fathers in the Philippines, particularly in those providing direct care: Lee T. Gettler, Thomas W. McDade, Alan B. Feranil, and Chris-

topher W. Kuzawa, "Longitudinal Evidence That Fatherhood Decreases Testosterone in Human Males," *Proceedings of the National Academy of Sciences* 108, no. 39 (2011): 16194–99. Also see Gray, McHale, and Carré, "A Review of Human Male Field Studies of Hormones and Behavioral Reproductive Effort."

This relationship between paternity status and T is not always found in men, as in Peter B. Gray, Jody Reece, Charlene Coore-Desai et al., "Testosterone and Jamaican Fathers," *Human Nature* 28, no. 2 (2017): 201–18.

198 T has been shown to rise in response to cries: Van Anders, Tolman, and Jainagaraj, "Examining How Infant Interactions Influence Men's Hormones, Affect, and Aggression."

198 apart from serving as a precursor to estrogen: All estrogens are derived from androgens, but as a precursor to estrogen, T itself has no direct action. For that to be the case it must act via the androgen receptor to alter cellular activity.

198 but the evidence implicating T is weak: Maurand Cappelletti and Kim Wallen, "Increasing Women's Sexual Desire: The Comparative Effectiveness of Estrogens and Androgens," *Hormones and Behavior* 78 (2016): 178–93; and Beverly G. Reed, Laurice Bou Nemer, and Bruce R. Carr, "Has Testosterone Passed the Test in Premenopausal Women with Low Libido? A Systematic Review," *International Journal of Women's Health* 8 (2016): 599.

199 if no relationship is detected between T and aspects of female sexuality: Ann Kathryn Korkidakis and Robert L. Reid, "Testosterone in Women: Measurement and Therapeutic Use," *Journal of Obstetrics and Gynaecology Canada* 39, no. 3 (2017): 124–30; and Laurence M. Demers, "Androgen Deficiency in Women; Role of Accurate Testosterone Measurements," *Maturitas* 67, no. 1 (2010): 39–45.

199 rates of low libido vary widely: Edward O. Laumann, Alfredo Nicolosi, Dale B. Glasser, Anthony Paik, Clive Gingell, E. Moreira, and Tianfu Wang, "Sexual Problems Among Women and Men Aged 40–80 Y: Prevalence and Correlates Identified in the Global Study of Sexual Attitudes and Behaviors," *International Journal of Impotence Research* 17, no. 1 (2005): 39–57. On the prevalence of low libido in women in the United States: Reed, Nemer, and Carr, "Has Testosterone Passed the Test in Premenopausal Women with Low Libido?" For the prevalence of hypoactive sexual desire (low sexual desire accompanied by distress), see: Shalender Bhasin and Rosemary Basson, "Sexual Dysfunction in Men and Women," in *Williams Textbook of Endocrinology*, 787 (Philadelphia: Elsevier Saunders, 2011).

199 Low libido is less common among younger women: Raymond C. Rosen, Jan L. Shifren, Brigitta U. Monz, Dawn M. Odom, Patricia A. Russo, and Catherine B. Johannes, "Epidemiology: Correlates of Sexually Related Personal Distress in Women with Low Sexual Desire," *Journal of Sexual Medicine* 6, no. 6 (June 2009): 1549–60.

199 there's little reason to think this is effective: Sheryl A. Kingsberg and Terri Woodard, "Female Sexual Dysfunction: Focus on Low Desire," *Obstetrics and*

Gynecology 125, no. 2 (2015): 477–86; and Cappelletti and Wallen, "Increasing Women's Sexual Desire."

199 And men may also be overprescribed testosterone: Richard G. Bribiescas, *How Men Age: What Evolution Reveals About Male Health and Mortality* (Princeton, NJ: Princeton University Press, 2018), 122.

200 libido does tend to increase, but so, too, do the masculinizing effects: Cappelletti and Wallen, "Increasing Women's Sexual Desire."

200 ovaries have for the most part closed up shop: Researchers disagree about whether the postmenopausal ovaries continue to produce meaningful levels of hormones, particularly androgens (low levels of androgens are continued to be made by the adrenal gland). See, e.g., Mario Vicente Giordano, Paula Almeida Galvão Ferreira, Luiz Augusto Giordano, Sandra Maria Garcia de Almeida, Vinícius Cestari do Amaral, Tommaso Simoncini, Edmund Chada Baracat, Mario Gáspare Giordano, and José Maria Soares Júnior, "How Long Is the Ovary Relevant for Synthesis of Steroids After Menopause?," *Gynecological Endocrinology* 34, no. 6 (2018): 536–39; and Fernand Labrie, "All Sex Steroids Are Made Intracellularly in Peripheral Tissues by the Mechanisms of Intracrinology After Menopause," *Journal of Steroid Biochemistry and Molecular Biology* 145 (2015): 133–38.

200 T levels drop by 30 to 50 percent after menopause: For postmenopausal steroid levels, see Robin Haring, Anke Hannemann, Ulrich John et al., "Age-Specific Reference Ranges for Serum Testosterone and Androstenedione Concentrations in Women Measured by Liquid Chromatography-Tandem Mass Spectrometry," *Journal of Clinical Endocrinology and Metabolism* 97, no. 2 (2012): 408–15.

200 Some studies on the efficacy of T to increase libido: Kingsberg and Woodard, "Female Sexual Dysfunction: Focus on Low Desire"; and Cappelletti and Wallen, "Increasing Women's Sexual Desire."

200 have normal sexual responses, desires, and orgasmic ability: Amy B. Wisniewski, Claude J. Migeon, Heino F. L. Meyer-Bahlburg et al., "Complete Androgen Insensitivity Syndrome: Long-Term Medical, Surgical, and Psychosexual Outcome," *Journal of Clinical Endocrinology and Metabolism* 85, no. 8 (2000): 2664–69.

200 Social and emotional factors: Bailey et al., "Effects of Gender and Sexual Orientation on Evolutionarily Relevant Aspects of Human Mating Psychology"; and Archer, "The Reality and Evolutionary Significance of Human Psychological Sex Differences."

200 Dopamine is upregulated in women in sexual contexts: Sheryl A. Kingsberg, Anita H. Clayton, and James G. Pfaus, "The Female Sexual Response: Current Models, Neurobiological Underpinnings and Agents Currently Approved or Under Investigation for the Treatment of Hypoactive Sexual Desire Disorder," *CNS Drugs* 29, no. 11 (2015): 915–33.

201 only clear example of exclusive homosexuality in a nonhuman animal: Simon LeVay, *Gay, Straight, and the Reason Why: The Science of Sexual Orientation* (Oxford: Oxford University Press, 2011), 119.

201   castrating them and mimicking the female sex hormone environment: P. Södersten, "Lordosis Behaviour in Male, Female and Androgenized Female Rats," *Journal of Endocrinology* 70, no. 3 (1976): 409–20.

202   Similar effects of sex hormones have been found in hundreds of studies: See LeVay, *Gay, Straight, and the Reason Why*, 31. Paradoxically, many of the masculinizing effects of prenatal androgens on the nervous system, at least in carnivores and rodents, occur via the actions of estrogen, which is produced at high levels prenatally via the conversion of testicular androgens. This is not the case in primates and humans. Several lines of evidence support this conclusion: first, men who cannot produce estrogen are fully masculinized (they show typical male behavior, interests, and sexual preferences); second, people with CAIS (who have XY sex chromosomes and who are sensitive to estrogen but not to androgens) are typically feminine.

202   lesbians are more likely to be attracted to male-dominated occupations: LeVay, *Gay, Straight, and the Reason Why*, 62; Lee Ellis, Malini Ratnasingam, and Mary Wheeler, "Gender, Sexual Orientation, and Occupational Interests: Evidence of Their Interrelatedness," *Personality and Individual Differences* 53, no. 1 (2012): 64–69; and Richard A. Lippa, "Sex Differences and Sexual Orientation Differences in Personality: Findings from the BBC Internet Survey," *Archives of Sexual Behavior* 37, no. 1 (2008): 173–87.

202   These gender-atypical feelings don't just pop up in adulthood: LeVay, *Gay, Straight, and the Reason Why*, 43–48; Michel Anteby, Carly Knight, and András Tilcsik, "There May Be Some Truth to the 'Gay Jobs' Stereotype," *LSE Business Review*, London School of Economics, January 18, 2016, https://blogs.lse.ac.uk/businessreview/2016/01/18/there-may-be-some-truth-to-the-gay-jobs-stereotype/; and András Tilcsik, Michel Anteby, and Carly R. Knight, "Concealable Stigma and Occupational Segregation: Toward a Theory of Gay and Lesbian Occupations," *Administrative Science Quarterly* 60, no. 3 (2015): 446–81.

203   girls who shun dresses and prefer rougher sports: J. Michael Bailey, Paul A. Vasey, Lisa M. Diamond, S. Marc Breedlove, Eric Vilain, and Marc Epprecht, "Sexual Orientation, Controversy, and Science," *Psychological Science in the Public Interest* 17, no. 2 (2016): 45–101.

203   This association between gender-atypical interests in childhood: Melissa Hines, "Prenatal Endocrine Influences on Sexual Orientation and on Sexually Differentiated Childhood Behavior," *Frontiers in Neuroendocrinology* 32, no. 2 (2011): 170–82.

203   Richard Green (then a psychiatrist at UCLA) and his team followed a group of boys: Richard Green, *The "Sissy Boy Syndrome" and the Development of Homosexuality* (New Haven, CT: Yale University Press, 1987), 12.

203   All the men in the gender-typical group were heterosexual: As cited in Hines, "Prenatal Endocrine Influences on Sexual Orientation and on Sexually Differentiated Childhood Behavior." See also Melissa Hines, Vickie Pasterski, Debra Spencer et al., "Prenatal Androgen Exposure Alters Girls' Responses to Information Indicating Gender-Appropriate Behaviour,"

Philosophical Transactions of the Royal Society B: Biological Sciences 371, no.
1688 (2016): 20150125; and Green, *The "Sissy Boy Syndrome,"* ch. 4.

204   both direct and indirect actions of T are at work: Sheri A. Berenbaum,
"Beyond Pink and Blue: The Complexity of Early Androgen Effects on Gender Development," *Child Development Perspectives* 12, no. 1 (2018): 58–64.

205   "not exclusively heterosexual": CAH and sexual orientation: Melissa Hines,
Mihaela Constantinescu, and Debra Spencer, "Early Androgen Exposure and
Human Gender Development," *Biology of Sex Differences* 6, no. 3 (2015); and
general population rate: LeVay, *Gay, Straight, and the Reason Why*, 8–9.

205   they grow up to be almost exclusively attracted to men: Martina Jürgensen,
Olaf Hiort, Paul-Martin Holterhus, and Ute Thyen, "Gender Role Behavior
in Children with XY Karyotype and Disorders of Sex Development," *Hormones and Behavior* 51, no. 3 (2007): 443–53; and Hines, Constantinescu,
and Spencer, "Early Androgen Exposure and Gender Development."

205   The genitalia differentiate during the first part: When T acts to masculinize
the nervous system, in non-primates it does so by first being converted to
estrogen inside nerve cells and acts through estrogen receptors in the nervous
system. (It masculinizes genitalia through direct interaction with androgen
receptors.) It also acts on the nervous system through direct androgen action
(through androgen receptors). In humans and nonhuman primates, direct
androgen action (through androgen receptors) appears crucial, without a
clear role for conversion to estrogen.

Evidence from nonhuman primates also suggests that different parts of
the brain may differentiate at slightly different times to masculinize different
types of behaviors: there may be one critical period for sexual behavior, and
one for competitive/aggressive behavior. This is elegantly demonstrated
in Goy's classic experiment on monkeys: R. W. Goy, F. B. Bercovitch, and
M. C. McBrair, "Behavioral Masculinization Is Independent of Genital Masculinization in Prenatally Androgenized Female Rhesus Macaques," *Hormones
and Behavior* 22, no. 4 (1988): 552–71.

205   we'd want to measure the amount of T in the fetal blood: Note: keep in mind
that in female fetuses, some T comes from the fetal adrenal gland and some
from maternal circulation. Most of the T in male fetuses is from the testicles.

206   can't be sure that we're measuring T at the right time: Dennis McFadden, "On
Possible Hormonal Mechanisms Affecting Sexual Orientation," *Archives of
Sexual Behavior* 46, no. 6 (2017): 1609–14.

207   A low ratio is the masculine profile: S. Marc Breedlove, "Minireview: Organizational Hypothesis: Instances of the Fingerpost," *Endocrinology* 151, no. 9
(2010): 4116–22.

207   To be confident we're seeing the effects of prenatal T on any particular trait:
For an overview of digit ratio and homosexuality: LeVay, *Gay, Straight, and
the Reason Why*, 71–74.

207   these make a big splash in the media: Cheryl M. McCormick and Justin M.
Carré, "Facing Off with the Phalangeal Phenomenon and Editorial Policies:

A Commentary on Swift-Gallant, Johnson, Di Rita and Breedlove (2020)," *Hormones and Behavior* 120 (2020): 104710.

207 little evidence from digit ratio studies that they had lower T prenatally: LeVay, *Gay, Straight, and the Reason Why*, 74.

207 to throw a large penis-wrench into the low-T theory: Anthony F. Bogaert and Scott Hershberger, "The Relation Between Sexual Orientation and Penile Size," *Archives of Sexual Behavior* 28, no. 3 (1999): 213–21. The authors are proponents of the T-theory and suggest other explanations. For skepticism about the study, see LeVay, *Gay, Straight, and the Reason Why*, 126.

207 T could affect sexual orientation in males: One possibility is that T levels are normal for most of prenatal development, but they are unusually high or low during a critical period when neural circuitry for sexual orientation is formed. Another possibility is that T levels are normal throughout but the response of relevant brain areas to T is different. And yet another is that genes related to sexual orientation are transcribed at different rates. Or perhaps T is not involved at all, and instead, what makes the difference are genes or gene expression. A very well-replicated finding is the "older brother effect": the likelihood of male homosexuality increases with a man's number of older brothers. This is probably due to previous male fetuses affecting the environment in the womb in ways that make homosexuality more likely for subsequent male fetuses. See Ray Blanchard, James M. Cantor, Anthony F. Bogaert, S. Marc Breedlove, and Lee Ellis, "Interaction of Fraternal Birth Order and Handedness in the Development of Male Homosexuality," *Hormones and Behavior* 49, no. 3 (2006): 405–14; and Charles E. Roselli, "Neurobiology of Gender Identity and Sexual Orientation," *Journal of Neuroendocrinology* 30, no. 7 (2018): e12562. And for a book-length treatment of the entire topic of the science of sexual orientation, see LeVay, *Gay, Straight, and the Reason Why*.

208 "What second date?": Andrew Sullivan, "#MeToo and the Taboo Topic of Nature," *New York Magazine*, January 19, 2018.

208 are much more likely to be sexual within committed monogamous relationships: Bailey et al., "Effects of Gender and Sexual Orientation on Evolutionarily Relevant Aspects of Human Mating Psychology."

208 want fewer sexual partners over their lifetime: A total of 30 to 50 percent of gay men are in a relationship, compared to 75 percent of lesbians. Bailey, *The Man Who Would Be Queen*, 87; and Christopher Carpenter and Gary J. Gates, "Gay and Lesbian Partnership: Evidence from California," *Demography* 45, no. 3 (2008): 573–90.

208 "Is male sexual aggression and horniness a function of patriarchy or testosterone?": Andrew Sullivan in Spencer Kornhaber, "Cruising in the Age of Consent," *Atlantic*, July 2019.

209 it's not a "gay" thing, it's a "man" thing: Bailey, *The Man Who Would Be Queen*, 87.

CHAPTER 9

211   "Everything I looked at, everything I touched turned to sex": "Testosterone: Act Two, Infinite Gent," *This American Life*, August 30, 2002, https://www .thisamericanlife.org/220/transcript.

211   "do not conform to what is typically associated with their sex assigned at birth": American Psychological Association, "Guidelines for Psychological Practice with Transgender and Gender Nonconforming People," *American Psychologist* 70, no. 9 (2015): 832–64.

A note on usage: The APA guidelines refer to biological sex as "sex assigned at birth." This term is used with increasing frequency, but I don't use it because it creates confusion. For one thing, it wrongly suggests that one's sex is the result of an arbitrary decision. Second, a person's sex may in rare cases differ from the sex they were assigned at birth. For example, people with DSDs, such as 5-ARD, might be assigned the female sex at birth, when they are actually male. These individuals may later identify either with their biological sex or with their sex assigned at birth. The two notions are distinct; both are useful.

211   1 in 250 (about 1 million) people identified as trans: Esther L. Meerwijk and Jae M. Sevelius, "Transgender Population Size in the United States: A Meta-Regression of Population-Based Probability Samples," *American Journal of Public Health* 107, no. 2 (2017): e1–e8; and Kenneth J. Zucker, "Epidemiology of Gender Dysphoria and Transgender Identity," *Sexual Health* 14, no. 5 (2017): 404–11. There are no reliable figures for the UK, but government estimates yield a roughly similar figure. See https://assets.publishing.service .gov.uk/government/uploads/system/uploads/attachment_data/file/721642 /GEO-LGBT-factsheet.pdf.

212   "gender dysphoria": American Psychological Association, "Guidelines for Psychological Practice with Transgender and Gender Nonconforming People," 2–3.

212   Some people feel so self-conscious about their bodies: These anxieties need not be equated with gender dysphoria; they're described to help readers imaginatively approximate what gender dysphoria feels like.

212   The transgender celebrity Jazz Jennings: Jeanette Jennings and Jazz Jennings, "Trans Teen Shares Her Story," *Pediatrics in Review* 37, no. 3 (2016): 99–100.

212   Gender dysphoria can also begin later: Kenneth J. Zucker, Anne A. Lawrence, and Baudewijntje P. C. Kreukels, "Gender Dysphoria in Adults," *Annual Review of Clinical Psychology* 12 (2016): 217–47; and K. J. Zucker, "Gender Identity Disorder in Children and Adolescents," *Annual Review of Clinical Psychology* 1 (2005): 467–92.

212   In children, the distress usually resolves: On the resolution of gender dysphoria ("desistance"), see Kenneth J. Zucker, "The Myth of Persistence: Response to 'A Critical Commentary on Follow-up Studies and "Desistance" Theories About Transgender and Gender Non-Conforming Children' by Temple Newhook et al. (2018)," *International Journal of Transgenderism* 19, no. 2 (2018): 231–45.

212 Often the transition will also be medical: American Society of Plastic Surgeons, "Gender Confirmation Surgeries," 2020, https://www.plasticsurgery .org/reconstructive-procedures/gender-confirmation-surgeries. For new research on options for cross-sex hormones, see India I. Pappas, Wendy Y. Craig, Lindsey V. Spratt, and Daniel I. Spratt, "Testosterone (T) and Estradiol (E2) Therapy Alone Can Suppress Gonadal Function in Transgender Patients," *Costas T. Lambrew Research Retreat 2020*, 47, https:// knowledgeconnection.mainehealth.org/lambrew-retreat-2020/47.

213 In England, for example, the number of young natal females: National Health Service (UK), "Referrals to the Gender Identity Development Service (GIDS) Level Off in 2018–19," Tavistock and Portman NHS Foundation Trust, June 28, 2019, https://tavistockandportman.nhs.uk/about-us/news /stories/referrals-gender-identity-development-service-gids-level-2018-19/.

213 Hormones and surgery for transgender people are booming: In the United States, coverage for transgender services, such as hormones and surgery, varies. For details of relevant policies from individual plans, see Human Rights Campaign, "Finding Insurance for Transgender-Related Healthcare," August 1, 2015, https://www.hrc.org/resources/finding-insurance-for-transgender -related-healthcare. In the UK, relevant services may be covered by the National Health Service (NHS), but one must qualify for the services and there may be a significant waiting period. For more information, see National Health Service, "Gender Dysphoria: Treatment," May 28, 2020, https://www .nhs.uk/conditions/gender-dysphoria/treatment/.

213 when testosterone is high: Gloria R. Mora and Virendra B. Mahesh, "Autoregulation of the Androgen Receptor at the Translational Level: Testosterone Induces Accumulation of Androgen Receptor mRNA in the Rat Ventral Prostate Polyribosomes," *Steroids* 64, no. 9 (1999): 587–91.

213 "he had the life-changing opportunity to transition from female to male and finally live life authentically": Buck Angel, "About," 2020, https://buckangel .com/pages/about-us.

215 direct the body's energy to be used to build up molecules and tissues: Hormones that build larger molecules out of smaller ones are called "anabolic," and those that break down molecules into smaller ones are called "catabolic."

215 T recruits a team of hormones: Peter T. Ellison, "Endocrinology, Energetics, and Human Life History: A Synthetic Model," *Hormones and Behavior* 91 (2017): 97–106.

216 The quality of a person's voice: Teresa L. D. Hardy, Jana M. Rieger, Kristopher Wells, and Carol A. Boliek, "Acoustic Predictors of Gender Attribution, Masculinity–Femininity, and Vocal Naturalness Ratings Amongst Transgender and Cisgender Speakers," *Journal of Voice* 34, no. 2 (2020): 300; Teresa L. D. Hardy, Carol A. Boliek, Daniel Aalto, Justin Lewicke, Kristopher Wells, and Jana M. Rieger, "Contributions of Voice and Nonverbal Communication to Perceived Masculinity-Femininity for Cisgender and Transgender Communicators," *Journal of Speech, Language, and Hearing Research* 63, no. 4 (2020):

931–47; and Adrienne B. Hancock, Julianne Krissinger, and Kelly Owen, "Voice Perceptions and Quality of Life of Transgender People," *Journal of Voice* 25, no. 5 (2011): 553–58.

216 A deep, strong voice is a potent signal of adult masculinity: For depth of voice and reproductive success (with review of attractiveness), see Coren L. Apicella, David R. Feinberg, and Frank W. Marlowe, "Voice Pitch Predicts Reproductive Success in Male Hunter-Gatherers," *Biology Letters* 3, no. 6 (2007): 682–84. For voice quality and sexual orientation, see Simon LeVay, *Gay, Straight, and the Reason Why: The Science of Sexual Orientation* (Oxford: Oxford University Press, 2011). Testosterone also predicts voice depth in men: James M. Dabbs Jr., and Alison Mallinger, "High Testosterone Levels Predict Low Voice Pitch Among Men," *Personality and Individual Differences* 27, no. 4 (1999): 801–4.

217 Although she can feminize it to some extent through voice therapy: See David Azul, Ulrika Nygren, Maria Södersten, and Christiane Neuschaefer-Rube, "Transmasculine People's Voice Function: A Review of the Currently Available Evidence," *Journal of Voice* 31, no. 2 (2017): 261.

217 male levels of T in puberty: Rahel M. Büttler, Jiska S. Peper, Eveline A. Crone, Eef G. W. Lentjes, Marinus A. Blankenstein, and Annemieke C. Heijboer, "Reference Values for Salivary Testosterone in Adolescent Boys and Girls Determined Using Isotope-Dilution Liquid-Chromatography Tandem Mass Spectrometry (Id-Lc–Ms/Ms)," *Clinica Chimica Acta* 456 (2016): 15–18; and David J. Handelsman, Angelica L. Hirschberg, and Stephane Bermon, "Circulating Testosterone as the Hormonal Basis of Sex Differences in Athletic Performance," *Endocrine Reviews* 39, no. 5 (2018): 803–29.

218 All this tubery: Eric P. Widmaier, Hershel Raff, and Kevin T. Strang, *Vander's Human Physiology: The Mechanisms of Body Function*, 14th ed. (New York: McGraw-Hill, 2015), 443.

218 tissues of the larynx are rich in androgen receptors: Scott-Robert Newman, John Butler, Elizabeth H. Hammond, and Steven D. Gray, "Preliminary Report on Hormone Receptors in the Human Vocal Fold," *Journal of Voice* 14, no. 1 (2000): 72–81. For a review of vocal tract development in adolescence, see Diana Markova, Louis Richer, Melissa Pangelinan, Deborah H. Schwartz, Gabriel Leonard, Michel Perron, G. Bruce Pike et al., "Age-and Sex-Related Variations in Vocal-Tract Morphology and Voice Acoustics During Adolescence," *Hormones and Behavior* 81 (2016): 84–96.

219 Other T actions also help masculinize the voice: Graham F. Welch, David M. Howard, and John Nix, *The Oxford Handbook of Singing* (Oxford: Oxford University Press, 2019), 24–25.

219 T also acts to lower the position of the larynx: W. T. Fitch and J. Giedd, "Morphology and Development of the Human Vocal Tract: A Study Using Magnetic Resonance Imaging," *Journal of the Acoustical Society of America* 106, no. 3 pt. 1 (1999): 1511–22. Red deer are among the rare mammalian species that also have a descended larynx, which appears to have evolved to intimidate rivals for mating competition: W. T. Fitch and D. Reby, "The Descended

Larynx Is Not Uniquely Human," *Proceedings of the Royal Society B: Biological Sciences* 268, no. 1477 (2001): 1669–75.

220   the length of the vocal folds is restricted: Azul et al., "Transmasculine People's Voice Function."

220   most trans men are satisfied with the vocal changes that T brings: Ulrika Nygren, Agneta Nordenskjöld, Stefan Arver, and Maria Södersten, "Effects on Voice Fundamental Frequency and Satisfaction with Voice in Trans Men During Testosterone Treatment—A Longitudinal Study," *Journal of Voice* 30, no. 6 (2016): 766, e24–e34.

220   a mistranslation of a Hebrew phrase: Wikipedia, "Adam's Apple," Etymology, retrieved August 15, 2020, https://en.wikipedia.org/wiki/Adam's _apple#Etymology.

220   A complicated series of events: Merriam-Webster, "Why Is It Called an 'Adam's Apple'? It's Not the Reason You Think," Merriam-Webster.com, Word History, https://www.merriam-webster.com/words-at-play/why-is-it-called -an-adams-apple-word-history.

221   angle at which the two cartilages join: Lee Coleman, Mark Zakowski, Julian A. Gold, and Sivam Ramanathan, "Functional Anatomy of the Airway," in Carin A. Hagberg, *Benumof and Hagberg's Airway Management*, 3rd ed., 3–20 (Philadelphia: W. B. Saunders, 2013).

221   the two plates of cartilage join at an angle of about 90 degrees: Neal S. Beckford, Dan Schaid, Stewart R. Rood, and Bruce Schanbacher, "Androgen Stimulation and Laryngeal Development," *Annals of Otology, Rhinology and Laryngology* 94, no. 6 (1985): 634–40.

221   "survival of the sweatiest": Bridget Alex, "Why Humans Lost Their Hair and Became Naked and Sweaty," *Discover*, January 7, 2019, https://www .discovermagazine.com/planet-earth/why-humans-lost-their-hair-and -became-naked-and-sweaty.

222   "most obvious biological marker of sexual maturity": Bridget Alex, "What Happened When Humans Became Hairless," *Discover*, August 13, 2019, https://www.discovermagazine.com/planet-earth/what-happened-when -humans-became-hairless; quote is from Barnaby Dixson, an anthropologist at the University of Queensland, Australia.

223   some trans men just can't grow the robust beards that they might like: E. J. Giltay and L. J. G. Gooren, "Effects of Sex Steroid Deprivation/Administration on Hair Growth and Skin Sebum Production in Transsexual Males and Females," *Journal of Clinical Endocrinology and Metabolism* 85, no. 8 (2000): 2913–21.

223   Blocking testosterone and increasing estrogen: Yi Gao, Toby Maurer, and Paradi Mirmirani, "Understanding and Addressing Hair Disorders in Transgender Individuals," *American Journal of Clinical Dermatology* 19, no. 4 (2018): 517–27.

224   the demand for "puberty blockers" has skyrocketed: Guido Giovanardi, "Buying Time or Arresting Development? The Dilemma of Administering Hormone Blockers in Trans Children and Adolescents," *Porto Biomedical Journal* 2, no. 5 (2017): 153–56.

224   "precocious puberty": Wassim Chemaitilly, Christine Trivin, Luis Adan, Valérie Gall, Christian Sainte-Rose, and Raja Brauner, "Central Precocious Puberty: Clinical and Laboratory Features," *Clinical Endocrinology* 54, no. 3 (2001): 289–94.

226   LH and FSH then travel to the gonads: See the figure on page 117.

227   about one pulse every sixty to ninety minutes: D. I. Spratt, L. S. O'Dea, D. Schoenfeld, J. Butler, P. N. Rao, and W. F. Crowley Jr., "Neuroendocrine-Gonadal Axis in Men: Frequent Sampling of LH, FSH, and Testosterone," *American Journal of Physiology* 254, no. 5, pt. 1 (1988): E658–66.

The frequency of GnRH pulses in men is relatively constant compared to that in women, in whom the frequency varies as a function of menstrual cycle phase. See Nancy Reame, Sue Ellyn Sauder, Robert P. Kelch, and John C. Marshall, "Pulsatile Gonadotropin Secretion During the Human Menstrual Cycle: Evidence for Altered Frequency of Gonadotropin-Releasing Hormone Secretion," *Journal of Clinical Endocrinology and Metabolism* 59, no. 2 (1984): 328–37.

227   second kind of period for brain organization: Sarah-Jayne Blakemore, Stephanie Burnett, and Ronald E. Dahl, "The Role of Puberty in the Developing Adolescent Brain," *Human Brain Mapping* 31, no. 6 (2010): 926–33.

228   delaying puberty may cause irreversible losses in bone strength: Caroline Salas-Humara, Gina M. Sequeira, Wilma Rossi, and Cherie Priya Dhar, "Gender Affirming Medical Care of Transgender Youth," *Current Problems in Pediatric and Adolescent Health Care* 49, no. 9 (2019): 100683.

228   blocking puberty may also reduce opportunities: Halting puberty has unknown effects on emotional development: Christopher Richards, Julie Maxwell, and Noel McCune, "Use of Puberty Blockers for Gender Dysphoria: A Momentous Step in the Dark," *Archives of Disease in Childhood* 104, no. 6 (2019): 611–12.

228   signing up for a lifetime of medical: To retain the effects of cross-sex hormones, they must be continued for life: Martin den Heijer, Alex Bakker, and Louis Gooren, "Long Term Hormonal Treatment for Transgender People," *BMJ* 359 (2017). Brief review of the role of hormones and surgery in transgender transitions: Jens U. Berli, Gail Knudson, Lin Fraser, Vin Tangpricha, Randi Ettner, Frederic M. Ettner, Joshua D. Safer et al., "What Surgeons Need to Know About Gender Confirmation Surgery When Providing Care for Transgender Individuals: A Review," *JAMA Surgery* 152, no. 4 (2017): 394–400.

228   around 95 percent of gender-dysphoric children who use puberty blockers go on to hormonally transition: We don't know why this is. One possibility is that people who choose to use blockers are those who would have hormonally transitioned in any event, blockers or no blockers. In other words, this group of people is highly motivated to transition and taking blockers was just a step along the path they were going to take no matter what. Another hypothesis is that taking blockers increases the likelihood of transition. This idea is supported by evidence showing that in the majority of even highly dysphoric adolescents, people come to feel more comfortable in their natal sex as puberty progresses. The developmental period between ten and thirteen appears to be

a particularly crucial time for adolescents to solidify gender identity, because it is during this period that the first glimmers of romantic feelings and sexual attraction may be experienced. For many, falling in love or coming to see oneself as a gay man or a lesbian, and not only accepting but embracing and enjoying one's sexually mature body, can decrease or even eliminate gender dysphoria. Similarly, when puberty only increases gender dysphoria, as was the case for Alan, who went through a feminizing puberty, the experience can intensify one's desire for and certainty about a gender transition. Unfortunately there is no test we can give that will tell us in advance which young people would be most likely to gain comfort with their natal sex and see dysphoria resolve, and which ones won't. See Richards, Maxwell, and McCune, "Use of Puberty Blockers for Gender Dysphoria."

228 the options for maintaining fertility: Timothy C. Lai, Rosalind McDougall, Debi Feldman, Charlotte V. Elder, and Ken C. Pang, "Fertility Counseling for Transgender Adolescents: A Review," *Journal of Adolescent Health* 66, no. 6 (2020): 658–65; Natnita Mattawanon, Jessica B. Spencer, David A. Schirmer, and Vin Tangpricha, "Fertility Preservation Options in Transgender People: A Review," *Reviews in Endocrine and Metabolic Disorders* 19, no. 3 (2018): 231–42; and D. Schlager, W. G. Lee, E. Williamson, R. Wafa, D. J. Ralph, and P. Sangster, "Fertility Preservation and Sperm Quality in Adolescent Transgender Patients Prior to Hormonal Treatment," *European Urology Open Science* 19 (2020): e533.

229 They may wish to transition back: Catherine Butler and Anna Hutchinson, "Debate: The Pressing Need for Research and Services for Gender Desisters/Detransitioners," *Child and Adolescent Mental Health* 25, no. 1 (2020): 45–47.

229 there are many detransitioners who are willing to share their stories: Van Slothouber, "(De) Trans Visibility: Moral Panic in Mainstream Media Reports on De/Retransition," *European Journal of English Studies* 24, no. 1 (2020): 89–99.

230 "I knew I was attracted to women": On sexual orientation in people with gender dsyphoria, see Anne A. Lawrence, "Sexual Orientation Versus Age of Onset as Bases for Typologies (Subtypes) for Gender Identity Disorder in Adolescents and Adults," *Archives of Sexual Behavior* 39, no. 2 (2010): 514-45.

230 "Each injection felt like an instant mood boost": Increases in T are associated with mood improvement: Michael Zitzmann, "Testosterone, Mood, Behaviour and Quality of Life," *Andrology* (July 13, 2020): 1–8.

232 "I became much more interested in men": Experiencing changes in patterns of sexual attraction after taking cross-sex hormones, as Stella and Kallisti did, is not uncommon: Matthias K. Auer, Johannes Fuss, Nina Höhne, Günter K. Stalla, and Caroline Sievers, "Transgender Transitioning and Change of Self-Reported Sexual Orientation," *PLoS One* 9, no. 10 (2014): e110016.

233 libido generally follows T: Mats Holmberg, Stefan Arver, and Cecilia Dhejne, "Supporting Sexuality and Improving Sexual Function in Transgender Persons," *Nature Reviews Urology* 16, no. 2 (2019): 121–39; and Michael S. Irwig, "Testosterone Treatment for Transgender (Trans) Men," in *The Plasticity of Sex*, ed. Marianne J. Legato, 137-57 (Amsterdam: Elsevier, 2020).

234 Male-to-female transitioners report: For interviews with transwomen about their changing sexuality, see Shoshana Rosenberg, P. J. Matt Tilley, and Julia Morgan, "'I Couldn't Imagine My Life Without It': Australian Trans Women's Experiences of Sexuality, Intimacy, and Gender-Affirming Hormone Therapy," *Sexuality and Culture* 23, no. 3 (2019): 962–77.

234 less research about emotional changes on and off T: Anger tends to increase and emotional expression (including crying) tends to decrease in transmen (on T); emotional expression and crying tend to increase in transwomen on estrogen (and blocking T). See Giovanna Motta, Chiara Crespi, Valentina Mineccia, Paolo Riccardo Brustio, Chiara Manieri, and Fabio Lanfranco, "Does Testosterone Treatment Increase Anger Expression in a Population of Transgender Men?," *Journal of Sexual Medicine* 15, no. 1 (2018): 94–101; and Justine Defreyne, Guy T'Sjoen, Walter Pierre Bouman, Nicola Brewin, and Jon Arcelus, "Prospective Evaluation of Self-Reported Aggression in Transgender Persons," *Journal of Sexual Medicine* 15, no. 5 (2018): 768–76.

236 emotional result of being on T for a year: Linden Crawford, "One Year on Testosterone," opinion, *New York Times*, June 18, 2020.

236 while women tend to let them bubble to the surface: Miranda A. L. Van Tilburg, Marielle L. Unterberg, and Ad J. J. M. Vingerhoets, "Crying During Adolescence: The Role of Gender, Menarche, and Empathy," *British Journal of Developmental Psychology* 20, no. 1 (2002): 77–87.

236 the same cannot be said for boys, whose tears seem to dry up: Johannes Fuss, Rainer Hellweg, Eva Van Caenegem, Peer Briken, Günter K. Stalla, Guy T'Sjoen, and Matthias K. Auer, "Cross-Sex Hormone Treatment in Male-to-Female Transsexual Persons Reduces Serum Brain-Derived Neurotrophic Factor (BDNF)," *European Neuropsychopharmacology* 25, no. 1 (2015): 95–99.

236 frequency of feeling and expressing anger doesn't differ much between males and females: John Archer, "The Reality and Evolutionary Significance of Human Psychological Sex Differences," *Biological Reviews* 94, no. 4 (2019): 1381–415.

237 Kallisti described a reduction in physical aggression: When studies have reported that aggression in men does change in response to T increases, it's usually in the predicted direction, is not a large effect, and depends on social and personality factors. Zitzmann, "Testosterone, Mood, Behaviour and Quality of Life."

237 there's no reason to think that T will turn a placid female-to-male transitioner into a hot-tempered Incredible Hulk: Anger does not typically increase with high doses of T in human males: R. Tricker, R. Casaburi, T. W. Storer, B. Clevenger, N. Berman, A. Shirazi, and S. Bhasin, "The Effects of Supraphysiological Doses of Testosterone on Angry Behavior in Healthy Eugonadal Men—A Clinical Research Center Study," *Journal of Clinical Endocrinology and Metabolism* 81, no. 10 (1996): 3754–58; nor does anger change with T suppression for medical reasons. Although human behavior is significantly impacted by hormones, relative to nonhuman animals we are somewhat liberated, with social, cognitive, cultural, and psychological factors making a larger contribution to the modulation of behavior. For this reason, large T changes in nonhuman

animals are likely to make a greater impact on behavior, as we saw with the Talapoin monkeys that received large doses of T and became more aggressive to the monkeys that were subordinate to them, as described in chapter 7.

237   childhood-onset gender dysphoria in natal females is strongly associated with sexual attraction to women: Kenneth J. Zucker, "Adolescents with Gender Dysphoria: Reflections on Some Contemporary Clinical and Research Issues," *Archives of Sexual Behavior* 48, 1983–1992 (2019): 1986.

CHAPTER 10

240   Miller revealed her true identity on an episode of *60 Minutes*: CBS News, "'Know My Name': Author and Sexual Assault Survivor Chanel Miller's Full *60 Minutes* Interview," August 9, 2020, https://www.cbsnews.com/news/chanel -miller-full-60-minutes-interview-know-my-name-author-brock-turner -sexual-assault-survivor-2020-08-09/.

241   Jonsson tripped Turner and pinned him down: Lindsey Bever, "The Swed-ish Stanford Students Who Rescued an Unconscious Sexual Assault Victim Speak Out," *Washington Post*, June 8, 2016; and Scott Herhold, "Thanking Two Stanford Students Who Subdued Campus Sex Assault Suspect," opin-ion, *Mercury News*, March 21, 2016, https://www.mercurynews.com/2016 /03/21/herhold-thanking-two-stanford-students-who-subdued-campus-sex -assault-suspect/.

241   "steep price to pay for 20 minutes of action": Elle Hunt, "'20 Minutes of Action': Father Defends Stanford Student Son Convicted of Sexual Assault," *Guardian*, June 5, 2016.

241   "You took away my worth . . .": Chanel Miller, *Know My Name: A Memoir* (New York: Viking, 2019), 343, 349.

242   Persky was recalled by California voters: Maggie Astor, "California Voters Remove Judge Aaron Persky, Who Gave a 6-Month Sentence for Sexual Assault," *New York Times*, June 6, 2018.

242   The sole female winner: Carnegie Hero Fund Commission, "15 Named Carnegie Heroes for Acts of Extraordinary Heroism," June 22, 2020, https:// www.carnegiehero.org/awardee_pr/15-named-carnegie-heroes-for-acts-of -extraordinary-heroism/.

243   take more physical risks than women: For "badass" women who excel in extreme sports, see Toby, "5 Most Badass Female Extreme Sports Athletes," Liftoff Adventure, March 12, 2019, https://liftoffadventure.com/most-badass -female-extreme-sports-athletes/.

For sex differences in impulsivity, risk-taking, and sensation seeking (men higher), see Marcus Roth, Jörg Schumacher, and Elmar Brähler, "Sensation Seeking in the Community: Sex, Age and Sociodemographic Comparisons on a Representative German Population Sample," *Personality and Individual Differences* 39, no. 7 (2005): 1261–71; Elizabeth P. Shulman, K. Paige Harden, Jason M. Chein, and Laurence Steinberg, "Sex Differences in the Developmental Trajectories of Impulse Control and Sensation-Seeking from Early

Adolescence to Early Adulthood," *Journal of Youth and Adolescence* 44, no. 1 (2015): 1–17; Marvin Zuckerman, Sybil B. Eysenck, and Hans J. Eysenck, "Sensation Seeking in England and America: Cross-Cultural, Age, and Sex Comparisons," *Journal of Consulting and Clinical Psychology* 46, no. 1 (1978): 139; and Catharine P. Cross, De-Laine M. Cyrenne, and Gillian R. Brown, "Sex Differences in Sensation-Seeking: A Meta-Analysis," *Scientific Reports* 3, no. 1 (2013): 1–5.

243   "thank you to the two men who saved me": Miller, *Know My Name*, 357.

243   accused Weinstein of sexual assault and harassment: Ronan Farrow, "From Aggressive Overtures to Sexual Assault: Harvey Weinstein's Accusers Tell Their Stories," *New Yorker*, October 10, 2017.

244   "asking them to look at your dick isn't a question": Louis C.K., "Louis C.K. Responds to Accusations: 'These Stories Are True,'" *New York Times*, November 10, 2017.

244   #MeToo is a movement that has made real progress: While the #metoo movement has been a powerful force in changing the culture around sexual assault and harassment, it has also sometimes overreached, and not all accused men have been found guilty. The journalist Emily Yoffe has documented examples of the personal and professional consequences: Emily Yoffe, "I'm Radioactive," *Reason Magazine*, October 2019.

244   "rates of rape and violence against wives and girlfriends have been sinking": Steven Pinker, *Enlightenment Now: The Case for Reason, Science, Humanism, and Progress* (New York: Penguin, 2018), 220–21.

245   which makes for differences among people: Robert Plomin, *Blueprint: How DNA Makes Us Who We Are* (Cambridge, MA: MIT Press, 2019), ix.

245   Without that gene, the body almost always develops as female: David C. Page, Rebecca Mosher, Elizabeth M. Simpson, Elizabeth M. C. Fisher, Graeme Mardon, Jonathan Pollack, Barbara McGillivray et al., "The Sex-Determining Region of the Human Y Chromosome Encodes a Finger Protein," *Cell* 51, no. 6 (1987): 1091–104.

246   It changes the way genes are expressed: Kristin R. Lamont and Donald J. Tindall, "Androgen Regulation of Gene Expression," *Advances in Cancer Research* 107 (2010): 137–62.

249   "Is it just a coincidence that this alleged surge in socialization comes at the same time": Steve Stewart-Williams, *The Ape That Understood the Universe: How the Mind and Culture Evolve* (Cambridge: Cambridge University Press, 2018), 109.

249   "What we know is that nurture becomes nature": Peggy Orenstein, as quoted in Isaac Chotiner, "Can Masculinity Be Redeemed?," *New Yorker*, January 20, 2020.

250   *Guardian* review of Cordelia Fine's book *Testosterone Rex*: Sarah Ditum, "Review: *Testosterone Rex* by Cordelia Fine: The Question of Men's and Women's Brains," *Guardian*, January 18, 2017.

251   "hormones make the man or woman, and we are what we secrete": Ditum, "Review: *Testosterone Rex* by Cordelia Fine."

253   "the idea that you were born this way and can't change who you are": Lynn Neary, "How 'Born This Way' Was Born: An LGBT Anthem's Pedigree,"

American Anthem, on *All Things Considered*, National Public Radio, January 30, 2019, https://www.npr.org/2019/01/30/687683804/lady-gaga-born-this-way-lgbt-american-anthem.

254 "Rather than rejecting biological differences, we must seek to reveal the nonsense": Sheri Berenbaum, "Biology: Born This Way?," *Science* 355, no. 6322 (2017): 254.

254 people are more forgiving: Matthew S. Lebowitz, "The Implications of Genetic and Other Biological Explanations for Thinking About Mental Disorders," *Hastings Center Report* 49 (2019): S82–S87.

254 "the steroids are what caused (Williams's) violent behavior": Kurt Greenbaum, "Steroid Defense Rejected, Jury Finds Suspect Guilty of Murder," *Sun Sentinel*, June 8, 1988, https://www.sun-sentinel.com/news/fl-xpm-1988-06-08-8802030649-story.html.

255 It's a general worry about whether biochemical causes of our actions excuse them: for an answer (and lively exploration of the question), see Robert M. Sapolsky, *Behave: The Biology of Humans at Our Best and Worst* (New York: Penguin, 2017), 580–613.

257 "Why Can't We Hate Men?": Suzanna Danuta Walters, "Why Can't We Hate Men?," *Washington Post*, June 8, 2018.

259 we have constructed fabulously sophisticated societies: Joseph Henrich, *The Secret of Our Success: How Culture Is Driving Human Evolution, Domesticating Our Species, and Making Us Smarter* (Princeton, NJ: Princeton University Press, 2017).

ACKNOWLEDGMENTS

I only have a few pages to thank everyone who helped me get this book written! It's a daunting task, since my instinct is to begin at conception. Please forgive me if I've left you out—it's because of space limits (in my brain).

Richard Wrangham took a risk on someone with no relevant training and gave me the opportunity to live in Uganda studying wild chimpanzees. That experience laid the foundation for this book. Having Richard as a friend and mentor has its trade-offs. For instance, you will feel that your knowledge, writing, and speech are all inadequate. You can nod along and pretend you know exactly what he's talking about, avoid him, or just up your game. I did all of the above, and am better for it. Richard, I can never thank you enough. If Richard is one of the most important ultimate reasons for this book, then Dan Lieberman is one of the most important proximate ones. Dan believed that writing a book was something that I not only should do, but could do. He did not give up on me when my first, then second, book proposals were rejected and I was convinced that I didn't have it in me. Thank you, Dan, for your endless harassment and support.

My literary agent, Max Brockman, had the good sense to reject those first two proposals. The longest thing I'd ever written was my dissertation, and I was naïve about all that is involved in getting a book proposal sold, then a book written, illustrated, edited, and published. Thank you, Max, and everyone at Brockman, Inc., for always being a pleasure to deal with, expertly handling all the legal and financial stuff and getting my book into the world.

My editor Maddie Jones at Holt patiently guided me through the writing process and helped me to clarify and organize my writing—not to mention being flexible when Covid made my writing situation less than ideal. Also, thank you to Gillian Blake (previously at Holt), Serena Jones at Holt, who saw me through the initial, and then final, stages of publishing, and Toby Lester, who did some masterful editing in the initial stages. Thanks to Anne McGuire for help with the references and notes. And to my two sensitivity readers: thank you for your keen attention to detail.

Dan Friedman, my professor and mentor from Antioch College, taught me how to think and write and introduced me to the joys of research. Dan: I strive to be as generous with my students as you were with me. Josephine Wilson: you sparked my interest in the biological basis of human behavior, and I will never forget those transformative moments in your class.

Thank you to my Ugandan field assistants John Barwogeza, Christopher Katongole, Francis Mugurusi, Donor Muhangyi, Christopher Muruuli, and Peter Tuhairwe, for collecting data, clearing the way, and teaching and protecting me.

Peter Ellison got me hooked on hormones and taught me much of what I know about the endocrine system and its relationship to human behavior. Learning and teaching about it all (with the help of Randy Nelson's excellent textbook *An Introduction to Behavioral Endocrinology*) has been one of the greatest pleasures of my life. Peter inspired me to always reach higher as a researcher and teacher and to respect the opposition. Steve Kosslyn took me into his lab, made mental rotation interesting, and got me to measure it and testosterone. Steve provided me with an environment that made researching and writing a dissertation far more fun than I'd imagined it could be. Peter and Steve: thank you for your mentorship and support.

Brian Hare and Chris Chabris: if I could have chosen people with whom I had to be stuck in a confined space for hundreds of hours, I never would have chosen either of you. So, lucky for me that space constraints rather than personal choice dictated decisions about my officemates in graduate school. I have the happiest of memories of my time with each of you, mostly of inappropriate jokes, TMI, and laughter but also of collaboration, productive debates, and lively conversations. You each helped me bring this book to life. For your friendship and support, and the most fun I've ever had at "work," thank you. Terrance Burnham,

Barbara Smith, Judith Flynn, and Matthew McIntyre: I think it's better that we didn't have cell phones back in the day, because there is no evidence. Thanks also to Jennifer Shephard, William Thompson, and Sam Moulton, and the entire Kosslyn lab. Susan Lipson: not everyone has the raw skills to work independently in the lab (that would be me). You have the patience of a saint, and if it were not for you, I would have had zero testosterone results to analyze for my dissertation research. And to Zarin Machanda: glad you didn't go far. You've been a great resource to me personally and professionally. A fount of evolutionary knowledge and gossip! Not sure I would have made it without you. And thank you to James Poolner, Mallory McCoy, and, especially, Meg Lynch, who kept everything running. I'll miss you.

This book could not have happened without the support of Joe Henrich, our department chair, and Logan McCarty, the director of Life Sciences. Thank you for your encouragement and making it possible for me to take the time I needed to focus on writing.

Neither Felix Byrne nor I knew what we were signing up for when I asked (more exactly, demanded) that he illustrate the book. Felix lives in a small village outside of Bath, UK, and communicating about the details of various graphs, glands, or pathways and how they should look was not always easy. Felix's talent and patience paid off with original works of art that bring T to life.

Having a beer by the river and a tour of the Cambridge colleges with Tim Clutton-Brock was a thrill. Tim, thank you for the day and for your contributions to my book. And especially for connecting me with Josephine Pemberton, who made the trip to Rum possible. Thank you to Sean and particularly Ali Morris, who hosted me on Rum and introduced me to Wisdom 11 and the rest of the stags, hinds, and their children. The sex and violence did not disappoint, and neither did the majesty of the place or the generosity and expertise of my hosts. And thank you to Fiona Guinness, who cooked for me a delicious fruit compote and let me pick her brain, which contains more knowledge about the deer than anyone on the planet.

Jenny, featured in chapter 1, is one of the bravest people I have ever known and taught me more than any book or paper could about DSDs. Thank you for your help. Alan, Kallisti, Sasha, and Stella, featured in chapter 9, all opened their lives up to me and my readers. I would not have been comfortable writing about trans and nonbinary experiences without their words. You made the "T in Transition" chapter come to life, and I am deeply grateful for your participation. It has been a pleasure to work with you. Daemon Fairless: chapter 7 starts off with a bang because of your personal story. Thank you for letting me use your words and experience.

Several people provided comments on chapters or sections of the book. I asked Steven Pinker whether he would be able to provide brief comments on one chapter, and a few days later I received several pages of detailed and incisive comments on the content and writing style of the entire book. For instance, he saved me from the embarrassment of "auxiliary hair," and needlessly modifying totally suitable adjectives. Steve, thank you for your kindness and generosity and for always following the evidence where it leads. My brother Mike Hooven, a mechanical engineer, also read the entire book and enlightened me, for example, about the fact that plumbing is not an "add-on" feature of a house. Richard Wrangham read several chapters and gave me detailed feedback. My wonderful students Chloe Ekhert and Anna Mazur not only provided some research but also regularly and helpfully disagreed with me about sensitive and controversial issues. I am also indebted to others who provided valuable comments: (in alphabetical order) J. Michael Bailey, Joyce Benenson, Andrew Berry, David Haig, David Handelsman, Fred Hooven, Tecumseh Fitch, Shawn Geniole, Peter Gray, Matthew Lebowitz, Martin Muller, Josephine Pemberton, and Jonna Vance.

Thank you to others who provided help in various forms: (in alphabetical order) Bridget Alex, Coren Apicella, Simon Baron-Cohen, Schuyler Bailar, Richard Bribiescas, Callie Burt, Jackie Byrne, Larry Cahill, Terry Capellini, Richard Clark, Doriane Coleman, Christine D'Ercole, Irv DeVore, Peter Eldredge, Melissa Emery Thompson, Frances Fuchs, Steve Gangestad, Dan Gilbert, Luke Glowacki, Abby Haas-Hooven, Molly Haas-Hooven, Ned Hall, Joanna Harper, Richard Holton, Maxwell Hooven, Ashley Judd, Sonya Kahlenberg, Karen Kramer, Rae Langton, Eleanor Lieberman, Andrew Light, Ally Love, Andrew McAfee, Barbara Natterson-Horowitz, David Page, David Pilbeam, Antonia Prescott, Sarah Richardson, Cody Rigsby, Diane Rosenfeld, Jane Rosenzweig, Elizabeth Ross, Maryellen Ruvolo, Mark Saia, Bill Segarra, Heather Shattuck-Heidorn, Jenn Sherman, Martin Surbeck, Eve Valera, Ian Wallace, David Watts, Christine Webb, Michael Wilson, Victoria Wobber, and Emily Yoffe.

It takes a village. Heidi Haas, you came through during the roughest of times. Special thanks to Susan, Dirk, Thomas, and Greta Koechner. We have a new home base in Germany! Andrea Abegglen and Barb, you gave me warm, beautiful, and welcoming places to focus and get this underway. Wendy Harrington, Matt and Edie Menard—you're a second family. Thank you to Katie Perkinson and Hugo and Maxwell Trappe; Amber, Marlon, and Conrad Kuzmick; Jane Rosenzweig; and David and Sam Barber for ALAP. Thanks to Katherine Sayn-Wittgenstein for sanity walks with the birds. Sean Kelly, thank you for the use of your secret office, and to Ned Hall and Barbara Popolow-Hall for nuts and turkeys.

My students! What a privilege it is to be able to teach, work with, and learn from you. Many of you have confided in me your struggles with being different, and some have shared your transitions with me, sometimes as they were underway. You've questioned my assumptions or language about sex, gender, and hormones. My life and thinking are richer because of my involvement with you. I have an amazing job.

Most of the students I teach at Harvard are high achievers who are mature, responsible, and organized, and they have been that way for much of their young lives. I wasn't like that as a young person (to put it mildly), and it's a daily struggle to make up for it. Jack and Marie Cort stepped in early on with love, a home base, a family, and encouragement, helping me get to a place where I could try.

Thank you to my dad, John G. Hooven. Martha and Steve Richardson: you took over when he left us, and you gave my boy grandparents. Frances and Naomi Fuchs, Mike, Fred, and John Hooven, I am the luckiest little sister. Aun, two lips. And to Griffin, I'm sorry! The writing really is done now. Thank you, sweetie, for putting up with me being unavailable and so much talk about gross hairy stuff.

Finally, thank you to my husband, Alex Byrne. I have the good fortune to be married to someone who is as interested in sex as I am. Alex served as my at-home editor, which led to some spirited discussions. Alex is a philosopher and a man—a tall one. I am neither of these things. But we still managed, about 60 percent of the time, to arrive at some semblance of an agreement about words or ideas that improved my writing. Alex, I know everyone says it, but there is really no way I could have done this without you.

INDEX

Page numbers in *italics* refer to illustrations.

ABOUT THE AUTHOR

CAROLE HOOVEN, PHD, is a lecturer and codirector of undergraduate studies in the Department of Human Evolutionary Biology at Harvard University. She earned her PhD at Harvard, studying sex differences and testosterone, and has taught there ever since. Hooven has received numerous teaching awards, and her popular Hormones and Behavior class was named one of the *Harvard Crimson*'s "Top Ten Tried and True."